Benefits of Bayesian Network Models

Systems Dependability Assessment Set

coordinated by
Jean-Francois Aubry

Volume 2

Benefits of Bayesian Network Models

Philippe Weber
Christophe Simon

ISTE

WILEY

First published 2016 in Great Britain and the United States by ISTE Ltd and John Wiley & Sons, Inc.

ISTE Ltd
27-37 St George's Road
London SW19 4EU
UK

www.iste.co.uk

John Wiley & Sons, Inc.
111 River Street
Hoboken, NJ 07030
USA

www.wiley.com

Library of Congress Control Number: 2016943665

British Library Cataloguing-in-Publication Data
A CIP record for this book is available from the British Library
ISBN 978-1-84821-992-2

Contents

Foreword by J.-F. Aubry

Systems Dependability Assessment is the title of a series of books, of which this is the third. The preface to the first series described the reasons why the authors embarked upon writing these books: in recent decades, they have made significant contributions to recent approaches to the predictive dependability of systems by considering concepts developed in other scientific fields but not yet applied to account of dependability. All these authors belong to the Automatic Control Research Center (CRAN, Centre de Recherches en Automatique de Nancy) of the University of Lorraine, France, a research laboratory whose activities are widely oriented towards the diagnosis, reliability, maintainability and safety of systems, which can be described in one word: dependability.

Assessment must be understood as the set of means, methods and tools to provide quantitative measures of dependability, and in these books we are interested in providing predictive measures by using probabilistic approaches. The first two books were dedicated to methods based on the frequentist knowledge of basic elements of a system and on models describing how the failure of this system depends on those of its components. These models are essentially of state-transition type, such as finite state automata and petri nets, and results were obtained by analytic or simulation approaches according to the level of complexity introduced, and data inputs for these models were probabilistic distributions of elementary events that were derived from strong feedback. This is within the approach called the frequentist or objective of probability theory.

The present book is different as it is based on Bayesian networks. Frequentist and Bayesian approaches to probabilities were both developed at about the same time in the 18th Century, but the first approach culminated in industrial development in the 19th Century, eclipsing the second. The original mathematical formulation of subjective probabilities was made almost simultaneously by Laplace and Bayes, the latter having given his name to the theorem of probability of causes. The Bayesian approach supposes an *a priori* knowledge, even if only approximate, of an event probability. From the knowledge of the *a priori* probabilities of the event and of its cause, the Bayes formula gives an *a posteriori* probability of the event, its likelihood function somehow describing the causal dependency. It is, in fact, a means of improving the knowledge of the event probability.

Bayes theorem can be materialized by a causal decision tree and extended to represent chained causality in a system. This is the base of Bayesian networks. With the development of computing technologies, the second half of the 20th Century saw the return of Bayesian methods, being an efficient tool to aid decision making in an uncertain environment. They provided solutions to problems such as climate change prediction or, more recently, the detection of spam in data communications.

Dependability assessment problems do not escape from Bayesian approaches. CRAN was among the first to promote these techniques, proposing methods and tools in association with industrial users and developers. Original works have been conducted, especially in the dependability of multi-state systems, integration of environmental and operating constraints in reliability computation (dynamic context) and interaction between the dependability and control of systems. The reader will find in this book a clear presentation of all these advances and I do not doubt that he/she can find a substantial benefit.

Jean-François AUBRY
Professor Emeritus
Lorraine University
Nancy, France

Foreword by L. Portinale

Probabilistic graphical models and Bayesian belief networks, in particular, have definitely become a reference formalism in dependability modeling and assessment. The graphical structure, together with the compact representation of the joint distribution of the system variables of interest, provides the reliability engineer with a powerful tool at both the modeling and analytical levels.

The dependency structure, induced by the graph component of the formalism, allows the modeler to make explicit a set of reasonable independence assumptions that may lead to huge simplification at the computational level, as well as with respect to the problem of probability elicitation, without compromising the suitability of the model produced to the actual real-world application.

Standard dependability models usually fit into two categories: 1) combinatorial models (as fault trees or reliability block diagrams) – they determine the occurrence of an undesired event through a combinatorial composition of sub-events; this class of model is very easy to analyze, but it cannot model situations involving complex dependencies among system components and sub-systems; 2) state–space models (such as Markov chains or petri nets) – they allow complex interactions among system parts to be modeled, but they may incur the “state-explosion” problem; this usually means that the analysis has to be performed by considering the cross-product of the system variables, producing a potentially huge number of states.

Bayesian networks and related models allow for efficient factorization of the set of system states, without the need for an explicit representation of the whole joint distribution; moreover it has the additional advantage of inference algorithms available for the analysis of any *a posteriori* situation of interest (i.e. evidence can be gathered by a monitoring system and fed into a dependability framework for fault detection and identification). Finally, when time is explicitly taken into account, models such as dynamic Bayesian networks result in a factored representation of a Markov chain, providing a framework with the modeling advantages of state–space models, without the drawbacks at the analytical level.

The present book, written by some of the most respected researchers and practitioners in the field, provides a comprehensive presentation and analysis of the probabilistic graphical model approach to dependability, providing a view of the different facets involved in real-world dependability applications: system reliability, maintenance and risk evaluation. The main objective is to devise a principled approach to the modeling of complex dependable systems, with the aim of supporting decisions in an uncertain and evolving setting. This supports and promotes Bayesian networks and probabilistic graphical models as some of the most relevant and important formalisms in modern dependability analysis.

Luigi PORTINALE
Computer Science Institute (DiSIT)
University of Eastern Piedmont
Alessandria, Italy

Acknowledgments

It is not easy to thank all those who have participated in and contributed to the research mentioned in this book.

I want to thank my scientific mentors Professor Benoît Iung and Professor Didier Theilliol, who have guided and supported my activity during all these years we have worked together. Special thanks to Christophe Simon, my co-author, my scientific partner and friend for all our joint contributions.

I thank my industry partners, Carole Duval (EDF), Paul Munteanu and Lionel Jouffe (Bayesia) for their trust and the various joint projects.

I cannot conclude the acknowledgments without extending my warmest thanks to my family, especially my wife Carole, who preserved the equilibrium in my life, and to my children Loïc and Manon, who have filled my family life with happiness.

Philippe WEBER
June 2016

Introduction

I.1. Problem statement

Since the beginning of the 20th Century, our perception of technological artifacts has continued to evolve and ranges from complicated systems to complex systems combining technical, human, organizational and environmental components. The first major accidents, such as the Flixborough blast [DEP 75], the Three Mile Island accident [KEM 79] or the toxic cloud at Seveso [SEV 82] that occurred at the end of the 1980s, have highlighted the role played by humans and their organizations in the failure of technical systems. The analysis of these accidents helps us to understand that the components are not independent. Thus, they have to be considered jointly to assess the true risk presented by our industrial systems.

Nowadays, the technological objects used are taken in their environment and are defined as complex socio-technical systems. The increase in complexity is due to the complexity emerging from the interdependence between technical, human, organizational and environmental components. A system is a set of elements interacting with each other with shared dependencies.

Faced with regulations, industrial systems are now required to have a high level of risk management. This level should be continuously demonstrated and proved [DE 12]. It is also necessary to handle socio-technical systems from a global point of view. To cope with this, dependability analysis and decision-making methods should be improved. Dependability analysis is primarily focused on technical aspects to assess the intrinsic safety of a system and should take into account human, organizational and environmental factors. Companies need good assessment tools to determine their requirements. They also need to anticipate future trends/developments to better manage or optimize the consequences of their activity on people, goods and the environment, as well as their social and societal impact.

To manage the industrial system, the engineers produced models during the lifecycle of a plant to predict its future functioning states. These models are extracted from our knowledge of the systems and our objective. There are several aims in dependability analysis. For instance, the following items are particular areas of interest in the Research Center for Automatic Control of Nancy (CRAN) in the University of Lorraine[1]:

– models that can assess the impacts of maintenance activities on the ability to maintain operational conditions and to aid decisions during maintenance;

– models that can assess the impacts of control and pilotage activities on wear, degradations (faults) or failures of all or part of the system. Thus, they satisfy the main goals, i.e. service quality, low-risk situations for users, staff, environment, etc.;

– models that help assess the efficiency of means to warrant an acceptable level of risk, whatever the operational constraints and environmental perturbations.

There are several domains of application of dependability. These domains can be associated with different aspects of system functioning such as management modes, governance, human factors, extreme events or rare events and their consequences on society, maintenance, control and supervision or risk reductions of socio-technical systems, etc.

1 BP 70239 – 54506 VANDOEUVRE, Cedex-France; http://www.cran.univ-lorraine.fr/.

Nevertheless, most engineers have neither the tools nor the methods to effectively understand the whole information set (knowledge and evidence) according to the operational constraints and disturbances that condition the functioning of socio-technical systems. This is the main paradigm regarding the management of socio-technical systems. The phenomena encountered are so complex, as a result of their heterogeneity and the number of nested mechanisms of different natures, that it is quite difficult to continuously meet the required objectives or levels of performance. Moreover, there are no exact analytical models that can describe all the phenomena encountered. It is also impossible to know all of the system states and to know and observe all of the component states at each point in time in order to determine the optimal decision. Engineers should bear in mind that all models are biased and partial. As a result, engineers need new methods to solve these modeling problems.

It is necessary to model systems and their components with a finite but unbounded set of states or performance levels, i.e. with multiple state systems. In addition, the component behaviors are conditioned by the operational constraints and environmental disturbances of the system. In such cases, dependability assessment becomes difficult because it should take into account the combining effects of dependent failures due to constraints, disturbances and the intrinsic multi-state nature of system components. This results in an increasing quantity of scenarios to model. It is cumbersome for the analyst and enforces bias and partiality.

Quantitative assessments are necessary to warrant the viability of systems and their performance regarding risk and dependability. It is thus necessary to handle an uncertain representation of the system to describe its functioning and dysfunctioning. This imperfect perception naturally leads to a probabilistic view of system states. The main difficulties are the integration of a huge amount of information to model industrial or socio-technical systems subject to a large set of interactions with its environment. To contribute to the solution of this modeling problem, this book shows the application of graph theoretic and probabilistic approaches using Bayesian networks (BN) in maintenance, risk analysis and management, as well as in control theory. Why is this choice made? In 2004, the Massachusetts Institute of Technology (MIT) published the rank of the next ten main revolutions in the industrial area. The use of BN was in fourth position.

In this book, the authors aim to formalize probabilistic graph model approaches like BN to solve different kinds of problems associated with dependability, maintenance and risk in complex systems. The book is oriented towards applications and the transfer of modeling methodologies to the industry and engineers. It does not focus on the algorithmic point of view, but on ways in which to build models for the dependability analysis of industrial systems. The book is inspired by some industrial problems the authors helped to solve with BN in several sectors.

The approach followed in the book is to link mathematical formalisms and their uses in industrial cases, industrial needs and the possibility of models. To do this, it is necessary to:

– generalize and homogenize the scientific paradigms from several industrial problems in an inductive way and then to specify solutions to applications in risk, reliability, maintenance and control;

– formalize and adapt methods to solve new industrial needs;

– promote these methods beyond the specialist community and adopt the standardization processes to integrate these methods.

In addition, two classes of problems need to be solved:

– the modeling of risk management, maintenance and reliability for socio-technical systems;

– the integration of knowledge of reliability in the control and diagnosis of automated systems.

The objectives of the book are to contribute to:

– modeling of complex systems to aid decisions in an uncertain context by proposing an efficient modeling method for the new challenges as socio-technical considerations;

– taking into account the propagation of uncertainties in complex system models, especially the uncertainty due to unknown future operational conditions;

– joint assessment of multi-sectorial risks by integrating organizational and human risks in the analyses;

– managing the aging of components, by anticipating the maintenance and safety actions with respect to aging;

– increasing the operational time with the objective of ensuring safety, risk management and quality even if faults or failures occur;

– allocating control efficiently to reduce the risk according to component criticality, health state and operational conditions.

To address these key points, the probabilistic framework and, particularly, BN formalism are used. BN are not completely accepted by the industry because the formalism does not handle a dependability-oriented semantic. In addition, the BN formalism is not proposed in usual standard modeling methods such as fault tree analysis[2], Markov chain[3], reliability block diagram[4], event tree analysis[5] and petri nets[6].

Currently, engineers mainly ask for the proof of correctness of models and results. The proof of computing correctness has been described in [PEA 88]. Thus, the remaining question concerns the correctness of the model built by the analyst. As BN offer a generic modeling framework, the subsequent question is how to attach a semantic to the model through a modeling methodology which is independent of the application. Moreover, the models obtained should achieve at least the same results and possibilities as well-recognized approaches. So, one objective of the book is to show and promote BN as a reference method.

2 IEC 61025 Fault tree analysis.

3 IEC 61165 Application of Markov techniques.

4 IEC 61078 Analysis techniques for dependability: reliability block diagram and Boolean methods.

5 IEC 62502 Analysis techniques for dependability: event tree analysis.

6 IEC 62551 Analysis techniques for dependability: petri net techniques.

I.2. Book structure

The scientific goal of this book is to formalize the probabilistic models of system functioning and dysfunctioning. The models built serve to assess the satisfaction of operational requirements and performances and the safety requirements through reliability and risk analyses. For this purpose, a system-centered model has to take into account the impact of the operational environment, control and maintenance strategies on the system health. These requirements are necessary because the engineer should maintain and optimize the quality of services performed by the industrial artifacts in use.

The probabilistic models are built by expertise and knowledge formalized from different types of standard modeling methods (FMEA[7], HAZOP[8], IDEF0[9] and structured analysis and design technique). The available data are considered as inadequate or are insufficient to build the model by learning algorithms. In the following, we propose the knowledge structuring principles that ensure a valid dependability model. The book gives examples of our scientific method to define the structure of probabilistic graphical models.

The book is organized as follows:

– Chapter 1 presents BN as a modeling formalism for system dependability. The modeling capability of a BN is explained by showing the relation between the joint probability over the system states and the system reliability. An academic system model illustrates the adaptability of the method to the dependability modeling problem of multi-state systems.

– Chapters 2 and 3 present BN as an easy, intuitive and smart formalism to modeling the structure function of complex and multi-state systems. These chapters introduce the principles for structuring a BN

7 IEC 60812 Analysis techniques for system reliability: procedure for failure mode and effects analysis (FMEA).

8 IEC 61882 Hazard and operability studies (HAZOP studies): application guide.

9 Icam DEFinition for Function Modeling. IEC 31320 Information technology: modeling languages (Part 1: syntax and semantics for IDEF0).

model in dependability. Several representations of the same problem are given and the chapters show the links between representations on the basis of the example given in Chapter 1. Some industrial applications are discussed to show the applicability of the modeling methods developed in the following PhD work and projects [LÉG 09, FAL 12, MED 11, ANR 11].

– Chapter 4 presents the integration of the temporal dimension in modeling based on dynamic Bayesian networks (DBN). This chapter explains different models of stochastic processes and describes the elegant representation provided by DBN for component reliability modeling. It ends with a discussion on modeling complexity and the difficulties in computation while merging several multi-state components into the model of the system. This allows us to understand the limitations of the approach and the work that still needs to be completed. DBN modeling is the solution adopted in Ben Salem's PhD thesis [BEN 08] to solve a diagnostic problem based on temporal sequences of events.

– Chapter 5 presents an integration of the DBN model of system reliability in a dynamic system control problem. The mutual influence of reliability on control and the control on reliability is exhibited. This chapter is based on work developed in Guenab and Khelassi's PhD theses [GUE 07, KHE 11].

– The conclusion focuses on the three major contributions of scientific research conducted at CRAN in the domain of probabilistic graphical model application in dependability.

PART 1

Bayesian Networks

1

Bayesian Networks: a Modeling Formalism for System Dependability

For complex systems, it is assumed that the system and the components have a finite number of states or operating levels. If the number of states is reduced to 2, then a binary hypothesis is used; otherwise the system and its components are multi-state. In this case, the evaluation of the reliability of the system becomes difficult, as it must take into account the effects of combinations of failures that are not independent of the multi-state nature of the system components. The result is the development of numerous modeling scenarios that become tedious for the analyst. In such cases, standard modeling procedures are insufficient, mainly due to their basis in Boolean logic or their need for (computationally expensive) randomized simulations.

As mentioned in [BOU 99], the modeling methods that come from artificial intelligence such as Bayesian networks (BN) can provide an effective support in control or maintenance areas, or in risk reduction for industrial systems. BN have powerful modeling and analysis capabilities. They provide a formal framework to handle or process probabilistic events by representing them using discrete random variables [PEA 88, JEN 96]. The relationships between them are represented by conditional probabilities. BN models are based on a powerful formalism of expressing complex dependence and independence between multi-state random variables. This formalism is, therefore, well suited to the representation of complex multi-state systems.

The application of BN in dependability analysis is a recent development. Their popularity has grown in the field of reliability analysis since the late 1990s [TOR 98, KAN 99]. According to [WEB 12b], up to 200 papers among 7,000 references have covered this field during the last decade.

Many of the significant papers in the application of BN to dependability have focused on demonstrating the equivalence between BN and conventional methods used in probabilistic risk assessment (PRA). For instance, Torres-Toledano's works [TOR 98] state the pro and cons of BN with respect to reliability block diagrams (RBD). Bobbio's works [BOB 01, BOB 03] explain how a fault tree can be transformed into a BN. In addition, [BOU 05b, BOU 05a, BOU 06, POR 10] describe the dynamic fault trees in dynamic BN and [WEB 02, WEB 03, WEB 06, BEN 06] link DBN with Markov chains.

At the same time, new modeling methods using BN modeling capabilities have emerged. One of the first papers [MAH 01] has proposed a methodology using BN for the assessment of infrastructure reliability. This methodology allows us to model two important characteristics of large structures: the modeling of multiple failure sequences and the correlations between the critical states at the component level.

The following papers [MUL 04, LÉG 09, MED 13, MED 15] have focused on a new methodology that is used to build BN models and to justify BN models for the reliability and risk analysis of complex systems. The publication of [BEN 13] has offered a method of construction for BN structure based on the cuts, to assess the performance of multi-state systems. These papers give clear motivation for using BN in dependability compared with conventional modeling methods.

Several survey papers [LAN 07, LAN 08, WEB 12b] have provided a good overview of BN applications in the domain of dependability.

Unfortunately, this modeling formalism is not yet fully accepted in industry. The French Institute of Risk Management (IMdR) has supported several projects to promote BN modeling. The IMdR P04-7 project [MUN 07] assesses the appropriateness of BN modeling for seven dependability problems posed by industrialists and the IMdR P09-2 project [GUY 09] addresses the issue of BN modeling validation.

1.1. Probabilistic graphical models: BN

In this section, the BN formalism is introduced as a probabilistic graphical model [PEA 88]. Mathematical objects are based on graph theory and probability theory. A BN represents a factorized model of a joint probability distribution of several discrete random variables. Graph theory provides the algorithms required to analyze graphical property. Probability theory brings a formalism to quantify the dependencies between variables by introducing conditional probability laws.

1.1.1. *BN: a formalism to model dependability*

The probabilistic graphical model (PGM) considered here is a directed acyclic graph (DAG). A DAG comprises nodes and directed edges. Nodes can be classified into two classes: parent nodes and child nodes. A parent node is a node with outgoing edges while a child node is a node with incoming edges. A parent node is called a root node if it has no incoming edges. A child node is a leaf node if it has no outgoing edge. Each parent node x in a graphical probabilistic model is assigned a marginal probability distribution $P(x)$ and each child node E is associated with a conditional probability $P(E|pa(E))$, where $pa(E)$ is the set of all parent nodes of E. For instance, in Figure 1.1 $pa(E_2) = x_2, x_3$ and $pa(E_1) = x_1$.

PGM is defined by the structure of the graph and the probabilistic parameters. According to the graph structure shown in Figure 1.1, the *a priori* probability laws are: $P(x_1)$, $P(x_2)$ and $P(x_3)$; while the conditional probabilities are $P(E_1|x_1)$, $P(E_2|x_2, x_3)$ and

$P(y|E_1, E_2)$. The conditional probabilities are defined by a conditional probability table (CPT) as a matrix giving the probability distribution of the variable with respect to the Cartesian product of its parent variable states. For instance, the conditional probability $P(y|E_1, E_2)$ is given in Table 1.1 for the $\{h_1^y, \ldots, h_n^y\}$ states of y, according to the $\{h_1^{E1}, \ldots, h_n^{E1}\}$ states of E_1 and the $\{h_1^{E2}, \ldots, h_n^{E2}\}$ states of E_2.

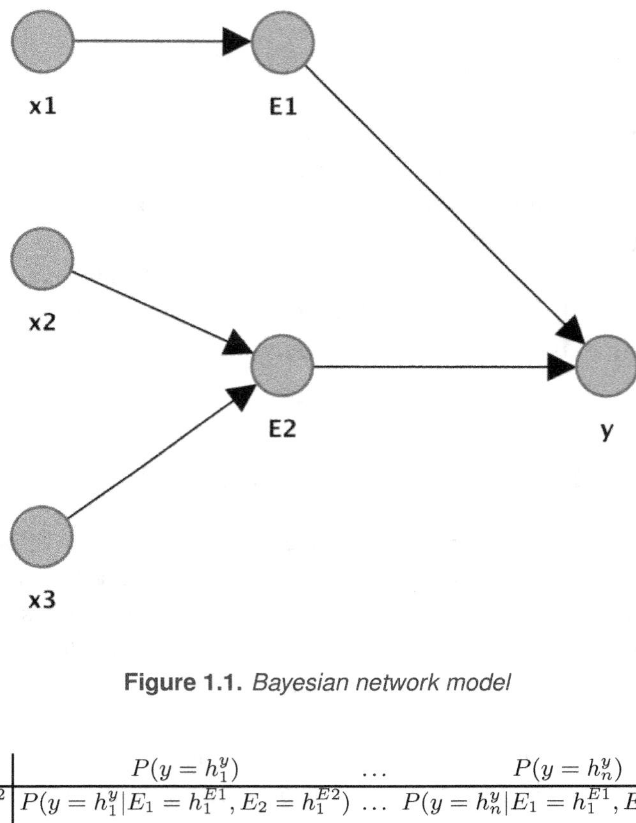

Figure 1.1. *Bayesian network model*

E_1	E_2	$P(y = h_1^y)$	$\ldots$	$P(y = h_n^y)$
h_1^{E1}	h_1^{E2}	$P(y = h_1^y \| E_1 = h_1^{E1}, E_2 = h_1^{E2})$	$\ldots$	$P(y = h_n^y \| E_1 = h_1^{E1}, E_2 = h_1^{E2})$
	$\ldots$			
	h_n^{E2}			
$\ldots$	$\ldots$		$\ldots$	
h_n^{E1}	h_1^{E2}			
	$\ldots$			
	h_n^{E2}	$P(y = h_1^y \| E_1 = h_n^{E1}, E_2 = h_n^{E2})$	$\ldots$	$P(y = h_n^y \| E_1 = h_n^{E1}, E_2 = h_n^{E2})$

Table 1.1. *Generic definition of a conditional probability table*

DEFINITION 1.1.– *A PGM represents the joint probability distribution of a set of random variables by using the conditional dependence/independence relationship between them. A DAG is used to represent the variables and the dependence relationships between them. Each variable is characterized by a probability distribution defined conditionally based on its parent's variables in the graph.*

1.1.2. *Inference mechanism*

Like several other modeling tools, BN are interesting for their graphical aspect. However, the probabilistic inference mechanism is more interesting and is the actual strength of the tool. Thanks to this inference, a BN is able to compute the marginal probability distribution of any variable according to:

– the realizations or measurements of observed variables (evidence);

– the likelihood regarding the state of certain variables;

– an *a priori* knowledge about the probability distribution of unobserved variables;

– the conditional probability distribution between variables.

The inference mechanisms are explained in [JEN 96, PEA 88] and are outside the scope of this book. Nevertheless, several inference mechanisms exist to compute the exact probabilities or the approximate probabilities for very complex systems. The inference algorithms are used to integrate new information in the model as soft or hard evidence. This information modeled as new observations on some variable states is a way to compute the impacts of situations on target variables. In maintenance or risk management, it is interesting to integrate specific situations or compute the impacts of some scenarios or maintenance actions. In all inference mechanisms, Bayes theorem is used to propagate the probabilities on the variables and to update the probabilities of all the variables given the observations of states or likelihoods of states.

In computer science, current research focuses mainly on inference efficiency to handle increasingly complex models and to increase the number of variables handled. For the exact inference, efficient algorithms use the BN structure to solve the non-deterministic polynomial-time-hard (NP-hard) problem to compute an *a posteriori* probability distribution of random variables [PEA 88, PEO 91, JEN 90, SHA 96, MAD 99, FAY 00, ALL 03]. The best known algorithms are based on the junction tree. For a detailed explanation, refer to [JEN 96, pp.76]. The newest algorithms attempt to reduce the memory requirements and to increase the computing speed to deal with larger models [JAE 02, WUI 12]. In dependability analyses, these abilities help model industrial-scale systems.

1.2. Reliability and joint probability distributions

BN are often of interest in the modeling of the dependability of systems. They allow a factorized model of the dependencies between the component states and the system states.

1.2.1. *Multi-state system example*

For a better understanding of BN applied to dependability modeling problems, let us analyze the multi-state system in Figure 1.2. This system contains three valves (V_1, V_2 and V_3). Its goal is to distribute a fluid. In this system, the components each have three states: a functioning state Ok and two disjoint failure states, i.e. remains closed Rc and remains open Ro [WEB 03]. Discrete variables (x_1, x_2 and x_3) model the states of the valves (V_1, V_2 and V_3). The probability of each of the states is given for each variable in Table 1.2.

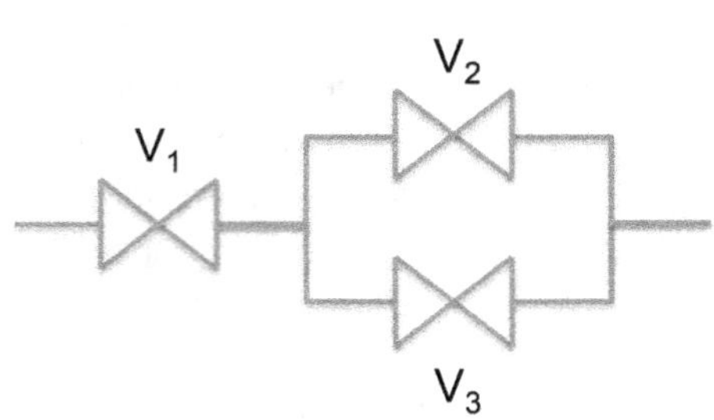

Figure 1.2. *Multi-state system with three valves*

	Ok	Rc	Ro
$P(x_1)$	0.31655	0.22782	0.45563
$P(x_2)$	0.19748	0.32095	0.48157
$P(x_3)$	0.14159	0.3678	0.49061

Table 1.2. *Probability distributions of component states*

1.2.2. *Joint distribution*

For any system, a probability is defined for each state corresponding to the Cartesian product between the states of each component and the system states to define the joint probability [SHA 96, p. 2]. The advantage of this representation is to show all the possible situations (working or failure). The main drawback is the size of the Cartesian product that increases rapidly and becomes excessive for the analyst, particularly in industrial-scale systems.

Tables 1.3 and 1.4 provide the application of the multi-state system with three valves. The joint probability is defined by $P(y, x_1, x_2, x_3)$, where y represents the system states and x_i represents the states of the components in the three-valve system. If the system is functioning, then $y = Ok$, otherwise $y = Hs$.

1.2.3. *Reliability computing*

The system reliability depends on the components' reliability (x_1, x_2 and x_3) and the relation between the system reliability, y, and the component states. This relation is the structure function. The joint probability distribution $P(y, x_1, x_2, x_3)$ can be computed on any structure function. In the case of the three-valve system, the reliability can be computed from the joint probability distribution. The reliability is then given by marginalization $P(y = Ok) = 0.345721859$, which is the sum of all state combinations, where the system state is Ok. Note that it is possible to compute all conditional probabilities from the joint probability distribution.

Scenarios	y	x_1	x_1	x_1	$P(y, x_1, x_2, x_3)$
1	Ok	Ok	Ok	Ok	0.008851116
2	Ok	Ok	Ok	Pf	0.022992022
3	Ok	Ok	Ok	Po	0.030669157
4	Ok	Ok	Pf	Ok	0.01438508
5	Ok	Ok	Pf	Pf	0
6	Ok	Ok	Pf	Po	0.049844368
7	Ok	Ok	Po	Ok	0.021584119
8	Ok	Ok	Po	Pf	0.056067794
9	Ok	Ok	Po	Po	0.074789071
10	Ok	Pf	Ok	Ok	0
11	Ok	Pf	Ok	Pf	0
12	Ok	Pf	Ok	Po	0
13	Ok	Pf	Pf	Ok	0
14	Ok	Pf	Pf	Pf	0
15	Ok	Pf	Pf	Po	0
16	Ok	Pf	Po	Ok	0
17	Ok	Pf	Po	Pf	0
18	Ok	Pf	Po	Po	0
19	Ok	Po	Ok	Ok	0.012739958
20	Ok	Po	Ok	Pf	0.033093839
21	Ok	Po	Ok	Po	0
22	Ok	Po	Pf	Ok	0.020705336
23	Ok	Po	Pf	Pf	0
24	Ok	Po	Pf	Po	0
25	Ok	Po	Po	Ok	0
26	Ok	Po	Po	Pf	0
27	Ok	Po	Po	Po	0

Table 1.3. *Joint probability distributions modeling the three-valve system, part 1*

1.2.4. *Factorization*

From the Cartesian product of states, a factorized version of the joint probability distribution can be computed by introducing conditional independence. Components x_1, x_2 and x_3 are state independent. Thus, the expression becomes $P(x_1, x_2, x_3) = P(x_1).\ P(x_2).P(x_3)$. Nevertheless, the functioning state of the system

$(y = Ok)$ depends on the state of components x_1, x_2 and x_3. The joint probability distribution can be rewritten in the following factorized form:

$$P(y, x_1, x_2, x_3) = P(x_1).P(x_2).P(x_3).P(y|x_1, x_2, x_3) \qquad [1.1]$$

Scenarios	y	x_1	x_1	x_1	$P(y, x_1, x_2, x_3)$
28	Hs	Ok	Ok	Ok	0
29	Hs	Ok	Ok	Pf	0
30	Hs	Ok	Ok	Po	0
31	Hs	Ok	Pf	Ok	0
32	Hs	Ok	Pf	Pf	0.037367275
33	Hs	Ok	Pf	Po	0
34	Hs	Ok	Po	Ok	0
35	Hs	Ok	Po	Pf	0
36	Hs	Ok	Po	Po	0
37	Hs	Pf	Ok	Ok	0.006370119
38	Hs	Pf	Ok	Pf	0.016547283
39	Hs	Pf	Ok	Po	0.022072492
40	Hs	Pf	Pf	Ok	0.010352895
41	Hs	Pf	Pf	Pf	0.026893105
42	Hs	Pf	Pf	Po	0.035872829
43	Hs	Pf	Po	Ok	0.01553402
44	Hs	Pf	Po	Pf	0.040351808
45	Hs	Pf	Po	Po	0.05382545
46	Hs	Po	Ok	Ok	0
47	Hs	Po	Ok	Pf	0
48	Hs	Po	Ok	Po	0.044144015
49	Hs	Po	Pf	Ok	0
50	Hs	Po	Pf	Pf	0.05378503
51	Hs	Po	Pf	Po	0.071744083
52	Hs	Po	Po	Ok	0.031067358
53	Hs	Po	Po	Pf	0.080701844
54	Hs	Po	Po	Po	0.107648537

Table 1.4. *Joint probability distributions modeling the three-valve system, part 2*

Equation [1.1] is the factorized form of the joint probability distribution $P(y, x_1, x_2, x_3)$. The conditional probability distribution $P(y|x_1, x_2, x_3)$, which is deterministic, remains unwieldy. Nevertheless, the conditional probability distribution may be factorized again by introducing intermediate variables, as done in fault trees. For instance, the system can be divided into two stages, as shown in Figure 1.3. Two variables E_1 and E_2 are introduced that characterize the states of the system stage. E_1 characterizes the possibility of controlling the flow in stage 1 and E_2 in stage 2.

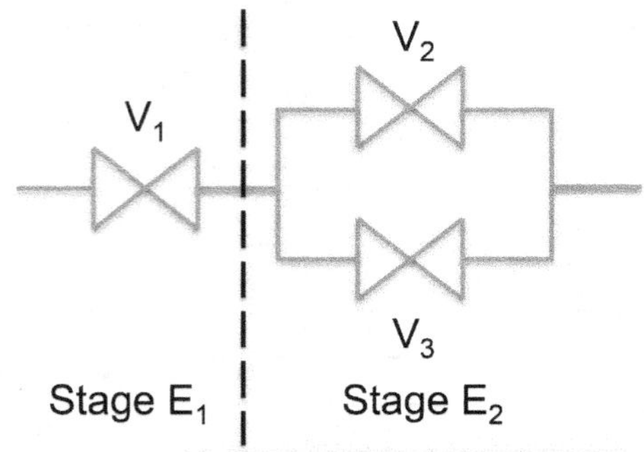

Figure 1.3. *Multi-state three-valve system with two stages*

To maintain all the information about the states, E_1 and E_2 are defined in three states:

– Ok, if it is possible to open or close the stage;

– if the stage failed and is stuck in the closed state and stops the flow;

– Ro, if the stage failed and is stuck in the open state and allows the flow.

Thus, equation [1.1] becomes:

$$P(y, x_1, x_2, x_3) = P(x_1).P(x_2).P(x_3).P(E_1|x_1).$$
$$P(E_2|x_2, x_3).P(y|E_1, E_2) \quad [1.2]$$

The probability distributions of the component states given in Table 1.2 are considered to be independent. They can be defined, thanks to the component lifetime or the end of mission, and estimated according to their reliability law (exponential, Weibull, etc.) or given by an expert.

The factorization of independent variables in the model helps simplify the model by a set of conditional probabilities whose size is considerably smaller than that of the joint probability. The conditional probability distributions are given in Tables 1.5, 1.6 and 1.7. Therefore, the BN is a graphical representation of this factorization of joint probability. By adding the graphical representation, the BN gives an easily interpretable model (Figure 1.1). The exact inference algorithms compute the marginal probability of all requested variables, for instance the probability distribution of y:

$$P(y = Ok) = 0.345721859 \quad [1.3]$$

$$P(y = Hs) = 0.654278141$$

x_1	$P(E_1 = Ok)$	$P(E_1 = Pf)$	$P(E_1 = Po)$
Ok	1	0	0
Rc	0	1	0
Ro	0	0	1

Table 1.5. *Probability distributions of E_1 states*

x_2	x_3	$P(E_2 = Ok)$	$P(E_2 = Pf)$	$P(E_2 = Po)$
Ok	Ok	1	0	0
	Rc	1	0	0
	Ro	0	0	1
Rc	Ok	1	0	0
	Rc	0	1	0
	Ro	0	0	1
Ro	Ok	0	0	1
	Rc	0	0	1
	Ro	0	0	1

Table 1.6. *Probability distributions of E_2 states*

E_1	E_2	$P(y = Ok)$	$P(y = Hs)$
Ok	Ok	1	0
	Rc	0	1
	Ro	1	0
Rc	Ok	0	1
	Rc	0	1
	Ro	0	1
Ro	Ok	1	0
	Pf	0	1
	Ro	0	1

Table 1.7. *Probability distributions of y states*

It should be noted that the probability distribution obtained by the BN model is the same as that obtained from the joint probability, such as $P(y = Ok)$.

1.3. Discussion and conclusion

The tables that define the conditional probabilities model the structure function of the system. This structure function is an equation that describes the relation between the component states and the system states. If the structure function is constant, then it implies that the conditional probability distribution is time independent. Defining the conditional probability using a table allows the modeling of any relations between the system states and the component states. If the relation is based on Boolean operation (AND, OR, etc.), then the CPT is deterministic, but more complex relations can be modeled. The reliability of the system is well modeled if the structure function is correctly modeled by the BN and if all scenarios are described. A CPT contains all the knowledge about the relation between the input states and the output states requested by the analysis.

In the classical case of binary state hypothesis, i.e. the system and its components can have two states $\{Ok, Hs\}$, the structure function is similar to a Boolean function. The CPT translates this Boolean relation. In this case, there is an exact correspondence between the BN model and a RBD when considering the working case or a fault tree when considering the failure case. Note that for our illustration, a non-binary function with three state components is deliberately chosen, to go beyond usual cases with RBD and fault tree and to exhibit part of the advantages of the BN model.

In our illustration as in all binary cases, there is no uncertainty between the combination of component states and system states. The probabilities of $P(y|x_1, x_2, x_3)$ are equal to 0 or 1. Therefore, CPT is deterministic. Note that this is not necessarily the case, for example, in a non-deterministic model, $P(y|x_1, x_2, x_3) \in [0, 1]$. This case models some situations where there is an uncertainty about the consequence of a component state combination, an uncertain function due to a human factor, an uncertain context, etc.

In this chapter, some of the main advantages of BN techniques have been discussed in an academic and industrial context. It is not necessary to know the joint probability of the system to find the BN model. The analyst can build the model gradually, but he should conduct his analysis with a semantic guide.

2

Bayesian Network: Modeling Formalism of the Structure Function of Boolean Systems

2.1. Introduction

One of the principle characteristics of modeling Boolean stucture function using BN is the ability to construct models from knowledge without technical expertise regarding computing algorithms. Nevertheless, this advantage can be a source of doubt about the computing results obtained from BN models. Formally, the numerical results are exact and the question of validity should concern only the quality of the model built by the analyst and/or the representativeness of data used to learn the parameters. Therefore, it is very important to use a structured modeling approach to obtain a model that better corresponds to reality.

From a practical point of view, there is often a lack of data to inform models in reliability estimation, risk analysis and maintenance optimization. It is often impossible to fully define the joint distribution defining all situations and their associated probabilities. As a result, modeling tools require the use of expert judgments to build structured models [CEL 06]. The BN modeling practiced in this book is presented in this spirit.

BN is a powerful modeling tool as it can combine knowledge of different kinds. This combination is allowed by the probabilistic representation and the combination of state of affairs. The model structure as well as the estimation of the model parameters can be built either automatically or manually from: data from feedback experiences; expert judgments based mainly on logical rules (not necessarily Boolean logic); equations; and databases of the system states or observations. By using objective or subjective probabilities, a BN can formalize causal relations or dependences/independences between variables. For instance, BN can model the effect of maintenance actions carried out by humans on a technical system (see [MED 11]) as well as the effect of defense barriers on risk analysis (see [LÉG 09]).

As previously discussed, BN are well-suited to modeling the structure function of system reliability. This modeling approach is based mainly on statistical knowledge and uses a combination of data and knowledge of qualitative causal relations to describe conditional dependencies between variables. The structure function is used in dependability analysis to model the propagation of failure events, degradation and alteration of the system [VIL 88, COC 97, COR 75, GER 00]. BN clearly helps in understanding the system behavior, thanks to the inference algorithm that propagates the observations (evidence) of the system and its components.

It is important to understand that more than one BN can model the same structure function. It depends on the way the factorization is done, exactly as with FT or RBD. The BN model structure will be different but the results are exactly the same because the same joint distribution is modeled.

Therefore, the PGM can be built by enumerating the (minimal) cut-sets and (minimal) tie-sets or based on the knowledge of system function. For the latter, a functional analysis like IDEF0 completed by dysfunctional analysis (FMEA or HAZOP) can help define the BN as mentioned in [WEB 06]. This approach reduces the tremendous task of enumerating cut-sets or tie-sets. To go beyond the BN, another way to build the model is to use the concept of object-oriented modeling with small bricks of knowledge. The reader can refer to SKOOB [ANR 11], a project funded by the French National Research Agency (ANR); an illustration of this methodology is described in [MED 11].

For an illustration of BN applied to dependability analysis, the following sections show the different BN models and their equivalence with usual models.

2.2. BN models in the Boolean case

Let us consider the binary state hypothesis. The BN model can be compared with FT, RBD, cut-sets and tie-sets. In this section, the risk analysis bowtie model is also introduced and a BN representation is given. This model has been used successfully in industrial applications [LÉG 09, FAL 12].

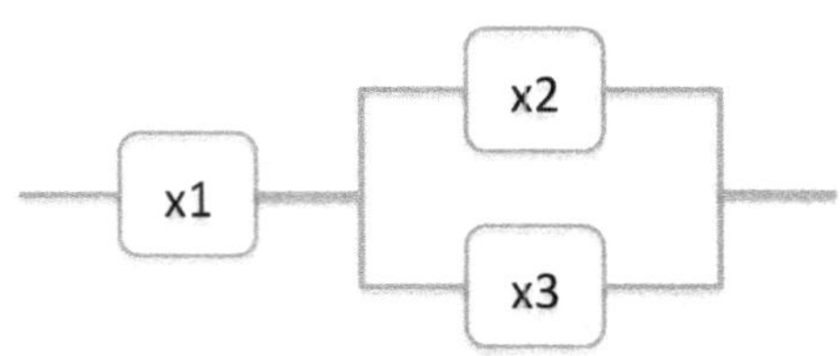

Figure 2.1. *RBD of the flow distribution system*

For the sake of illustration, this section focuses on the flow distribution system: the three-valve system, given in Chapter 1 in Figure 1.2. The RBD of this system is illustrated in Figure 2.1. The mission of the system is to distribute the flow. Contrary to Chapter 1, the binary state hypothesis is made. Thus, the components have two states: $x_i = 0$ if the valve i is working and allows the flow to go through the valve; and $x_i = 1$ if the valve i does not allow the flow to go through the valve – the valve is then considered broken. The probability distributions of x_i are given in Table 2.1. The system is modeled by the variable y: $y = 0$ if the system accomplishes its mission; and $y = 1$ if the system is unable to accomplish its mission.

	$P(x_i = 0)$	$P(x_i = 1)$
x_1	0.77218	0.22782
x_2	0.67905	0.32095
x_3	0.6322	0.3678

Table 2.1. *Probability distribution of valves' states*

2.2.1. *BN model from cut-sets*

The cut-sets represent the malfunction scenarios of the system. Based on the RBD illustrated in Figure 2.1, three cut-sets can be isolated:

$$\begin{aligned} C_1 &= \{x_1\} \\ C_2 &= \{x_2, x_3\} \\ C_3 &= \{x_1, x_2, x_3\} \end{aligned} \qquad [2.1]$$

The system works if no cut-set occurs or conversely the system fails if at least one of the cut-sets occurs. This property is due to the Boolean nature of the components. The BN model of reliability based on these three cut-sets is shown in Figure 2.2.

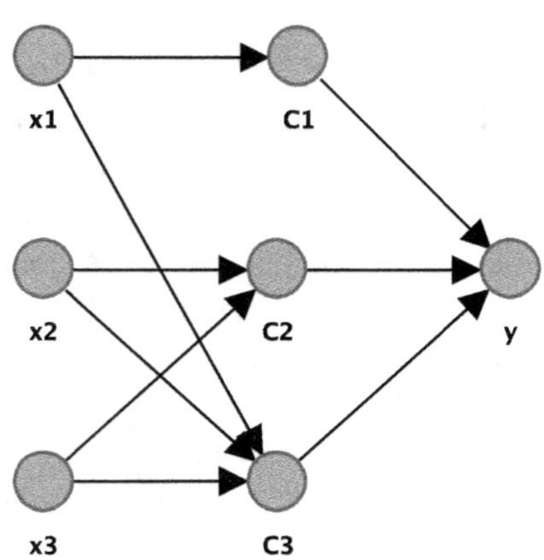

Figure 2.2. *BN model for three cut-sets*

Let us consider the probability distributions of the components' states given in Table 2.1 and the conditional probability tables of each of the cut-sets based on a deterministic equation. The CPT of cut-set C_2 is given in Table 2.2 and the CPT for y in Table 2.3 for the sake of illustration.

x_2	x_3	$P(C_i = 1)$	$P(C_i = 0)$
0	0	0	1
0	1	0	1
1	0	0	1
1	1	1	0

Table 2.2. *CPT of cut-set C_2*

C_1	C_2	C_3	$P(y = 0)$	$P(y = 1)$
0	0	0	0	1
0	0	1	0	1
0	1	0	0	1
0	1	1	0	1
1	0	0	0	1
1	0	1	0	1
1	1	0	0	1
1	1	1	1	0

Table 2.3. *CPT of y based on cut-sets*

The dependability engineer knows that it is efficient to compute the reliability directly with minimal cut-sets, i.e. cut-sets that do not include other cut-sets. Thus, equation [2.2] becomes equation [2.3] and consequently the BN model is reduced, as shown in Figure 2.3.

$$C_1 = \{x_1\} \quad [2.2]$$

$$C_2 = \{x_2, x_3\} \quad [2.3]$$

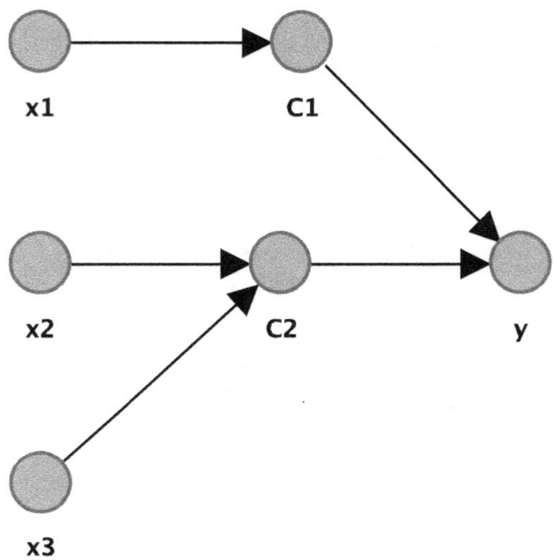

Figure 2.3. *BN model of two minimal cut-sets*

With BN, the two models give exactly the same correct results, whereas the first has one more cut-set. C_3 is, in fact, the combination of the two other cut-sets. When or C_2 occurs then C_3 occurs. With BN, the dependencies between cut-sets are taken into account during the inference and, hence, the results are correct. This is the strength of BN and their inference algorithms. Conditional dependencies are taken into account by inference algorithms to compute the exact results. Thus, the difference is in the computational cost, which is optimal when the model is based on minimal cut-sets.

2.2.2. *BN model from tie-sets*

Let us consider the problem from the point of view of success. Three tie-sets can be extracted from Figure 2.1, as given by equation [2.3]. The BN model (Figure 2.4) is given by:

$$\begin{aligned} L_1 &= \{x_1, x_2\} \\ L_2 &= \{x_1, x_3\} \\ L_3 &= \{x_1, x_2, x_3\} \end{aligned} \qquad [2.4]$$

The reader can note directly that none of the tie-sets are minimal because $L_1 \cup L_2 = L_3$. Nevertheless, as in the case of cut-sets, the inference mechanism will work properly and give the correct result. Let us consider the two minimal tie-sets L_1 and L_2, for computing the probability distribution of the system states. The deterministic CPT of tie-set L_1, according to the states of x_1 and x_2, are given in Table 2.4. The CPT for L_2 is given in Table 2.5. The combination of the two minimal tie-sets is enough to compute the probability distribution of y (Table 2.7), thanks to its CPT table (see Table 2.6).

x_1	x_2	$P(L_1 = 0)$	$P(L_1 = 1)$
0	0	1	0
	1	0	1
1	0	0	1
	1	0	1

Table 2.4. *CPT of* L_1

x_1	x_3	$P(L_2 = 0)$	$P(L_2 = 1)$
0	0	1	0
	1	0	1
1	0	0	1
	1	0	1

Table 2.5. *CPT of* L_2

L_1	L_2	$P(y=0)$	$P(y=1)$
0	0	1	0
	1	1	0
1	0	1	0
	1	0	1

Table 2.6. *CPT of $y|L_1, L_2$*

y	P(y)
0	0.681027695
1	0.318972305

Table 2.7. *Probability distribution on y state*

The logical behavior of the system failures induces deterministic CPT, which are equivalent to Boolean gates. The BN are modeling Boolean equations by probabilities equal to 0 or 1.

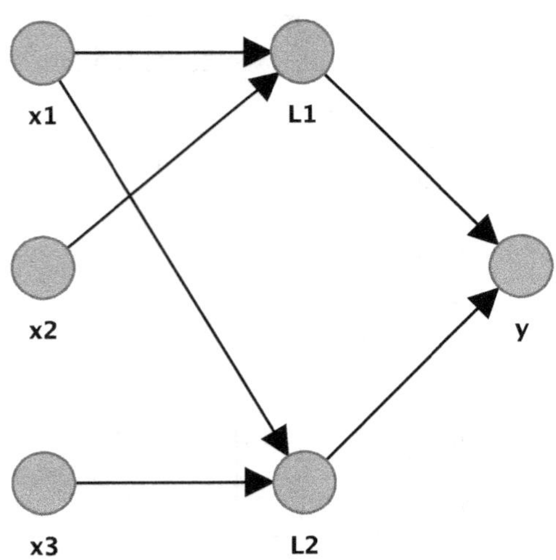

Figure 2.4. *BN modeling the two minimal tie-sets*

2.2.3. *BN model from a top-down approach*

In the case of large systems, the enumeration of all functioning or dysfunctioning scenarios is cumbersome. To solve this problem, the FT modeling is based on a descending approach. Starting from a top event that characterizes the undesired event, the analysis goes down the tree by the definition of intermediate events identified as direct causes of upper events, until elementary events are obtained. For example, Figure 2.5 shows the FT of the flow distribution system, which is obviously quite simple.

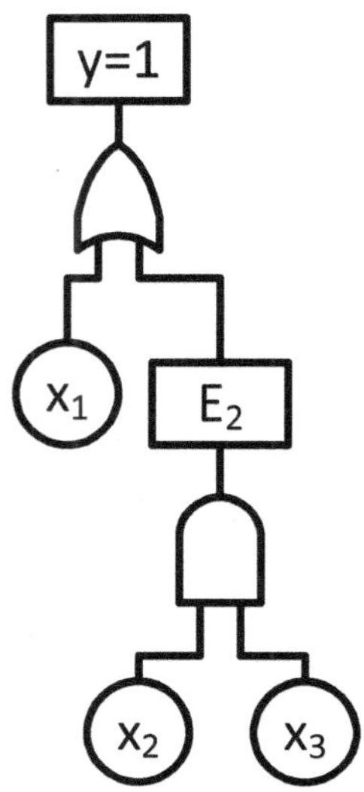

Figure 2.5. *FT of the flow distribution system*

If a FT is available, it is very simple to translate it into a BN by simple mapping. As shown earlier for tie-sets and cut-sets, deterministic CPT can map Boolean relations between variables with logical operators: AND and OR. Figure 2.6 shows the mapping result of the FT shown in Figure 2.5 into a BN. For each event in the FT, a variable is defined in the BN.

For instance, the AND gate in Figure 2.5 is such that $E_2 = x_2 \wedge x_3$, i.e. E_2 occurs if $x_2 = 1$ and $x_3 = 1$. The CPT of the BN is given in

Table 2.8. The OR gate in Figure 2.5 is such that $y = x_1 \vee E_2$, i.e. $y = 1$ for the failure of the system if $x_1 = 1$ or $E_2 = 1$. The CPT of the BN is defined in Table 2.9.

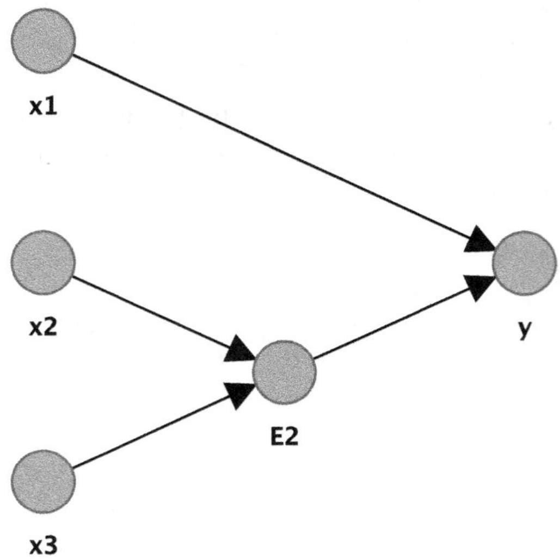

Figure 2.6. *BN model of the FT of Figure 2.5*

By the inference mechanism, the BN computes the probability distribution of y, which is the same as given in Table 2.7.

x_2	x_3	$P(E_2 = 0)$	$P(E_2 = 1)$
0	0	1	0
	1	1	0
1	0	1	0
	1	0	1

Table 2.8. *CPT of a Boolean AND*

2.2.4. *BN model of a bowtie*

In risk analysis, to assess the impacts of an undesired event, an event tree (ET) is added to the FT, resulting in a bowtie model. As it is quite easy to model a FT by an equivalent BN, an ET is modeled by an equivalent BN. Therefore, a BN model of a bowtie model can be obtained easily, as shown in Figure 2.7. Usually, in risk analysis, the goal is to place and assess barriers to prevent or protect from the undesired event. This modeling technique has been applied in several industrial cases [LÉG 09, DUV 12, KHA 13]. In Figure 2.7, variables E_2 and Ip_1 model the impact of barriers B_1 and B_2 on the propagation of the malfunctioning of the system defined by $y = 1$.

x_1	E_2	$P(y=0)$	$P(y=1)$
0	0	1	0
	1	0	1
1	0	0	1
	1	0	1

Table 2.9. *CPT of a Boolean OR*

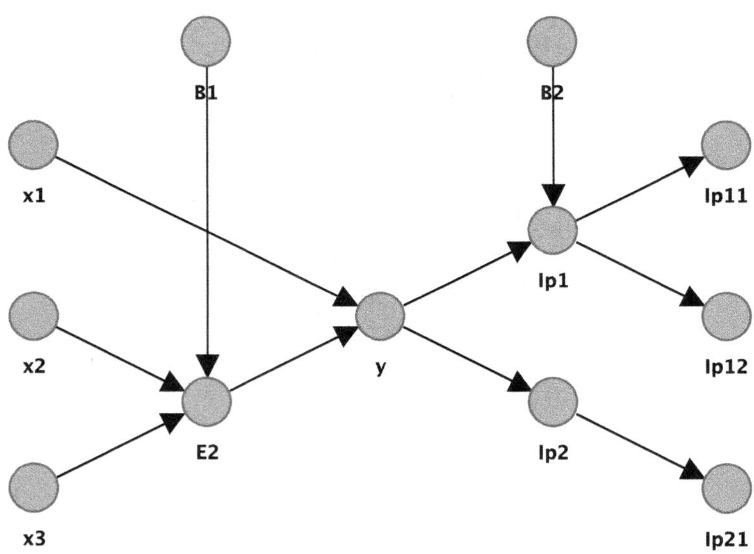

Figure 2.7. *BN model of a bowtie and its barriers*

In the proposed model, barriers act as an inhibitor of the causal influence between parent and child variables. In Figure 2.7, variables

B_i represent the efficiency of the barriers. $B_i = 0$ represents an efficient barrier that stops all the propagation of an undesired event. The CPT of the variable in the bowtie is given in Table 2.10. The relation between E_2, x_2 and x_3 is given by the FT in Figure 2.5 and CPT Table 2.8. The variable B_1 represents the inhibitor effect of the barrier. If $B_1 = 0$, the barrier stops the event propagation and $E_2 = 0$. If $B_1 = 1$ the barrier is not efficient, and the propagation is made as in the FT by the equation $E_2 = x_2 \wedge x_3$ exactly as without the barrier. The CPT of Ip_1 inhibited by B_2 is given in Table 2.11.

The model of barrier efficiency can be formalized as a graph combining all the factors of efficiency losses as a BN structure perpendicular to the bowtie and mapping on the FT as proposed in [LÉG 09, FAL 12]. [LÉG 09, FAL 12] have also defined the way to estimate the barrier efficiency from expert elicitations, taking into account human, organizational and environmental factors.

B_1	x_2	x_3	$P(E_2 = 0)$	$P(E_2 = 1)$
0	0	0	1	0
0	0	1	1	0
0	1	0	1	0
0	1	1	1	0
1	0	0	1	0
1	0	1	1	0
1	1	0	1	0
1	1	1	0	1

Table 2.10. *CPT of the inhibitor of $E_2 = x_2 \wedge x_3$ by B_1 in a bowtie model*

B_2	y	$P(Ip_1 = 0)$	$P(Ip_1 = 1)$
0	0	1	0
0	1	1	0
1	0	1	0
1	1	0	1

Table 2.11. *CPT of the inhibitor of Ip_1 by B_2 in a bowtie model*

2.3. Standard Boolean gates CPT

All Boolean gates can be modeled by a BN (OR, AND, Koon, Exclusive OR, etc.). It is sufficient to directly map the Boolean equation inside the CPT [SIM 07, SIM 08].

An n-component system that functions (or works) if and only if at least k of the n components work is called a k-out-of-n:G system. An n-component system that fails if and only if at least k of the n components fail is called a k-out-of-n:F system. Both parallel and series systems are special cases of the k-out-of-n system. A series system is equivalent to a 1-out-of-n:F system and an n-out-of-n:G system, while a parallel system is equivalent to an n-out-of-n:F system and a 1-out-of-n:G system.

Let us define the CPT of a 2-out-of-3:G system, with the components x_1, x_2 and x_3. The BN structure is shown in Figure 2.8. The system is functioning, $y = 0$, if at least two components are available; $x_i = 0$ and $x_j = 0$, with $i \neq j$ and $i, j \in \{1, 2, 3\}$. The CPT of y is defined in Table 2.12.

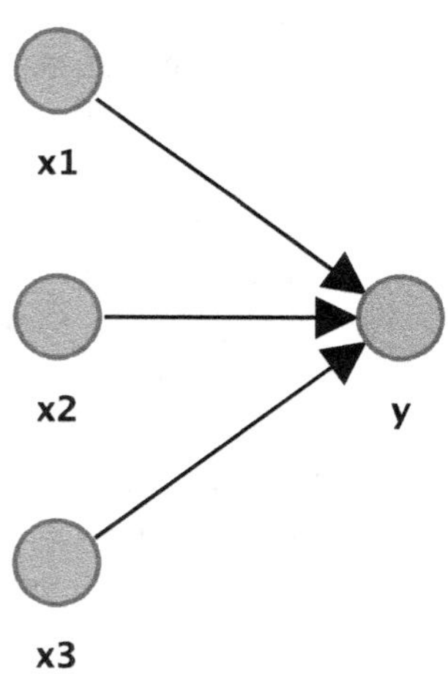

Figure 2.8. *BN model of the 2-out-of-3:G system*

Unlike FT or RBD, BN can integrate topological constraints, for instance the linear or circular consecutive-koon system. Such systems cannot be modeled by FT or RBD because of the independence hypothesis of events. The BN solves this problem by computing conditional independence and gives a systematic modeling method [WEB 10, WEB 11].

x_1	x_2	x_3	$P(y=0)$	$P(y=1)$
0	0	0	1	0
0	0	1	1	0
0	1	0	1	0
0	1	1	0	1
1	0	0	1	0
1	0	1	0	1
1	1	0	0	1
1	1	1	0	1

Table 2.12. *CPT of a Boolean 2-out-of-3:G system*

Consecutive-koon systems have attracted considerable attention since they were first proposed by Kontoleon in 1980 [KON 80]. A consecutive-koon system can be classified according to the linear or circular arrangement of its components and the functioning or malfunctioning principle. Thus, four types of k-out-of-n can be enumerated: linear consecutive-koon:F, linear consecutive-koon:G, circular consecutive-koon:F and circular consecutive-koon:G. A consecutive-koon:F system consists of a set of n ordered components that compose a chain such that the system fails if at least k consecutive components fail. A consecutive koon:G system is a chain of n components such that the system works if at least k consecutive components work. An illustration of these specific structures can be found in telecommunication systems with n relay stations that can be modeled as a linear consecutive-koon:G system if the signal transmitted from each station is strong enough to reach the next k stations. An oil pipeline system for transporting oil from point to point with n spaced pump stations is another example of a linear-consecutive-koon system. A closed recurring water supply system with n water pumps in a thermo-electric plant is a good example of a circular system. The system ensures its mission if each pump is powerful enough to pump water and steam to the next k consecutive pumps [YAM 03].

Let us define the BN model of a linear consecutive-2-out-of-5:G system and a circular consecutive-2-out-of-5:G system with the components x_1, x_2, x_3, x_4 and x_5. The BN structures are given in Figure 2.9 and Figure 2.10. The system is functioning, $y = 0$, if at least two consecutive components are available, $x_i = 0$ and $x_{i+1} = 0$, with $i \in \{1, 2, 3, 4, 5\}$. The factorization given by the variable C_i, based on AND gates, allows us to simplify the CPT of y, which is then only based on an OR gate. The CPT of the C_i is defined in Table 2.13 and the table of y is given in Table 2.14 for the linear consecutive-2-out-of-5:G system and in Table 2.15 for the circular consecutive-2-out-of-5:G system.

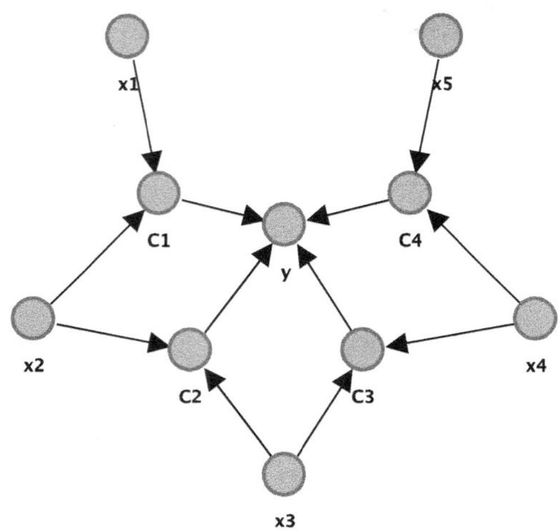

Figure 2.9. *BN model of the linear consecutive-2-out-of-5:G system*

2.4. Non-deterministic CPT

The binary state hypothesis is usually made while dealing with reliability or dependability analysis, as done in previous sections. Then, Boolean logic (OR, AND, XOR, NOT, etc.) defines the failure scenarios that lead to the undesired event as described by FT or equivalent representations. There is also the possibility of translating algebraic relations. In these situations, BN include deterministic CPT, i.e. conditional probabilities only in $0, 1$.

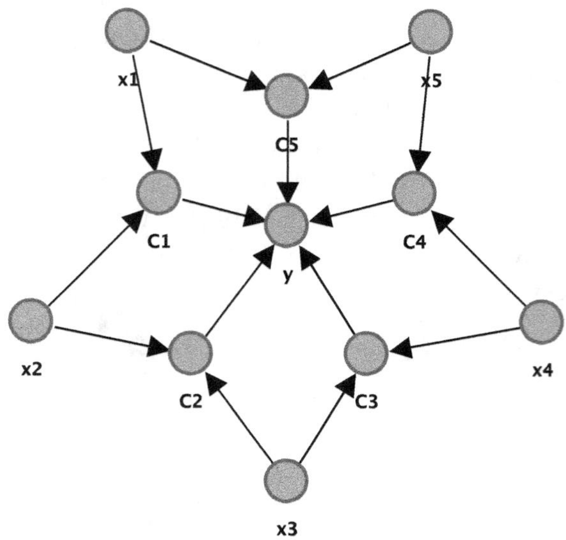

Figure 2.10. *BN model of the circular consecutive-2-out-of-5:G system*

x_i	x_{i+1}	$P(C_i = 0)$	$P(C_i = 1)$
0	0	1	0
	1	0	1
1	0	0	1
	1	0	1

Table 2.13. *CPT of the C_i variables*

Nevertheless, BN models are able to consider non-deterministic CPT. Non-deterministic CPT is defined by conditional probabilities in $]0, 1[$. It means that the occurrence of a cause cannot produce the consequence at all. If the CPT is built by the database, then the non-deterministic CPT arrives when the occurrences of some parent states do not produce the same occurrence of the child state. When the CPT is built by an analyst, the non-deterministic CPT translates the analyst's inability to define a strict logical relation between the parents and the considered variable. The problem occurs if the expert is unsure about the relation or if he suspects that other non-modeled elements influence the variable considered.

C_1	C_2	C_3	C_4	$P(y=0)$	$P(y=1)$
0	0	0	0	1	0
0	0	0	1	1	0
0	0	1	0	1	0
0	0	1	1	1	0
0	1	0	0	1	0
0	1	0	1	1	0
0	1	1	0	1	0
0	1	1	1	1	0
1	0	0	0	1	0
1	0	0	1	1	0
1	0	1	0	1	0
1	0	1	1	1	0
1	1	0	0	1	0
1	1	0	1	1	0
1	1	1	0	1	0
1	1	1	1	0	1

Table 2.14. *CPT of y in a linear consecutive-2-out-of-5:G*

When the CPT is defined from a large dataset, a learning algorithm solves the problem easily. Three principal algorithms exist: counting [HEC 96, KRA 98], expectation-maximization [LAU 95] and gradient descent [BIN 97]. The previous case considers a general case where all conditional probabilities are estimated in the interval $[0, 1]$. The problem of learning CPT arises when there is a small dataset [ONI 01]. This case is usually too incommodious to be defined by an expert. If there is a known relation between parents and the variable considered, this relation may simplify the expert's work. Henrion [HEN 89] talks about independence of the causal influence (ICI) models based on the assumption of ICI of the parents. This assumption leads to the number of parameters needed to build the CPT, proportional to the number of parents. A further distinction is made between:

– noisy ICI models;

– leaky ICI models (i.e. an extension of the formers);

– probabilistic ICI.

C_1	C_2	C_3	C_4	C_5	$P(y=0)$	$P(y=1)$
0	0	0	0	0	1	0
0	0	0	0	1	1	0
0	0	0	1	0	1	0
0	0	0	1	1	1	0
0	0	1	0	0	1	0
0	0	1	0	1	1	0
0	0	1	1	0	1	0
0	0	1	1	1	1	0
0	1	0	0	0	1	0
0	1	0	0	1	1	0
0	1	0	1	0	1	0
0	1	0	1	1	1	0
0	1	1	0	0	1	0
0	1	1	0	1	1	0
0	1	1	1	0	1	0
0	1	1	1	1	1	0
1	0	0	0	0	1	0
1	0	0	0	1	1	0
1	0	0	1	0	1	0
1	0	0	1	1	1	0
1	0	1	0	0	1	0
1	0	1	0	1	1	0
1	0	1	1	0	1	0
1	0	1	1	1	1	0
1	1	0	0	0	1	0
1	1	0	0	1	1	0
1	1	0	1	0	1	0
1	1	0	1	1	1	0
1	1	1	0	0	1	0
1	1	1	0	1	1	0
1	1	1	1	0	1	0
1	1	1	1	1	0	1

Table 2.15. *CPT of y in a linear consecutive-2-out-of-5:G*

The Noisy-OR structure was introduced by Pearl [PEA 88] to reduce the elicitation effort in building the BN probabilistic model that combines probability and logic. This structure is an extension of the OR structure. Let us consider a binary variable y with its binary parent variables x_i (Figure 2.11). Variables y and x_i can be either true 1 or false 0. X_T is defined as the set of x_i that are true and X_F is defined as the set of x_i that are false. The Pearl's idea is to associate with each x_i a "link probability" p_i such that $0 \leq p_i \leq 1$. This probability p_i corresponds to the probability that y is true if x_i is true. It illustrates the fact that the causal dependency between x_i and y can be inhibited (or imperfect). This link probability p_i is defined as follows:

$$p_i = P(y = \{1\}|x_i = \{1\}, X_{j,j \neq i} = \{0\}) \qquad [2.5]$$

With this proposal, only n parameters are sufficient to completely define the CPT of y. Then, it is easy to define the probability that y is true given x_i and the probability that y is false given x_i with the following relations:

$$P(y = 1|x_i) = 1 - \prod_{x_i \in X_T} (1 - p_i) \qquad [2.6]$$

$$P(y = 0|x_i) = \prod_{x_i \in X_T} (1 - p_i) \qquad [2.7]$$

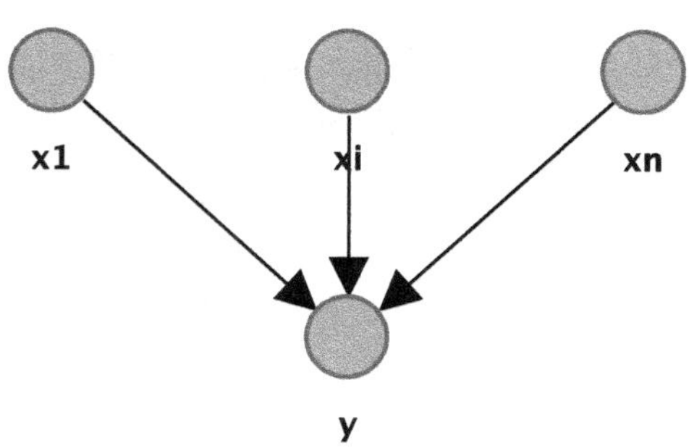

Figure 2.11. *Noisy-OR structures*

The Noisy-OR (N-OR) structure implies that $y = 0$ with a probability equal to 1 if all its parent variables $x_i = 0$. However, in many cases this is an unrealistic assumption. It is difficult to capture all the causes of y in several situations (e.g. for reliability to define all the failure modes of a component). Therefore, Henrion [HEN 89] proposed an extension of the N-OR structure called Leaky Noisy-OR (LN-OR) by introducing a new parameter called "leak probability". The leak probability corresponds to other unknown parents that may affect y. These parents are modeled by using the variable L with a link probability $p_i = l$ (Figure 2.12). Let l be the leak probability, such that $0 \leq l \leq 1$, which is defined as follows:

$$l = P(y = 1 | x_i = 0, \forall i = 1..n) \quad [2.8]$$

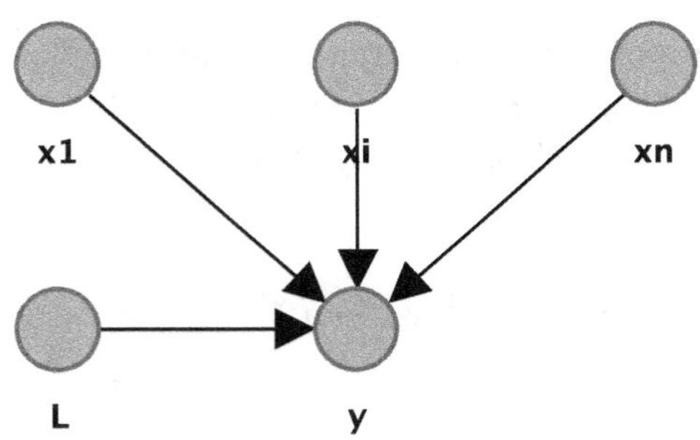

Figure 2.12. *Leaky Noisy-OR structures*

Henrion [HEN 89] proposed the following parameterization of the LN-OR structure with $n + 1$ parameters:

$$P(y = 1 | x_i) = 1 - (1 - l) \times \prod_{x_i \in X_T} (\frac{1 - p_i}{1 - l}) \quad [2.9]$$

$$P(y = 0 | x_i) = (1 - l) \times \prod_{x_i \in X_T} (\frac{1 - p_i}{1 - l}) \quad [2.10]$$

Diez [DIE 93] gave another parameterization of the LN-OR structure as follows:

$$P(y = 1|x_i) = 1 - (1 - l) \times \prod_{x_i \in X_T} (1 - p_i) \quad [2.11]$$

$$P(y = 0|x_i) = (1 - l) \times \prod_{x_i \in X_T} (1 - p_i) \quad [2.12]$$

These two parameterizations are mathematically equivalent, but the difference is related to the question the experts are asked to elicit knowledge. Henrion's parameterization led to the following question: "What is the probability that y is true given that x_i is true and all other modeled variables are false?" Diez's parameterization is supported by the following question: "What is the probability that y is true given that x_i is true and all other modeled and non-modeled variables are false?" In Henrion's parameterization, experts have to consider a combined influence of x_i and the leak on y. In Diez's parameterization, experts have to consider the link between x_i and y with the leak being absent.

The parameters defined earlier can be uncertain too. For example, the influence of a parent variable on a child variable, but also the number of non-modeled parent variables or the state of parent variables cannot be well known. In risk analyses, experts often use intervals to assess different variables. For example, intervals can be used to assess the state of a parent variable or the influence of a parent variable on the child variable. It is, therefore, necessary to take into account uncertainties relating to the link probabilities, the leak probability and the state of parent variables x_i. Scientific contributions already exist in relation to the problem of uncertainty in logical structures such as N-OR or LN-OR. Srinivas [SRI 93] and Diez [DIE 93] proposed an extension of the N-OR structure for non-Boolean variables. Antonucci [ANT 11] developed an imprecise LN-OR structure with uncertainty on the link probabilities that can be extended to uncertainty on the leak probability with Diez's parameterization.

2.5. Industrial applications

Over the past few years, several modeling approaches have dealt with a global view of risks, accounting for technical aspects while being immersed in human, organizational and environmental contexts. [PAT 96] developed the system–action–management (SAM) approach and [SVE 02] highlighted the importance of considering different actors in the risk analysis of an industrial system through a graphical representation of causal flow of accidents (AcciMap). Moreover, [PAP 03] developed the I-Risk method that can account for both technical and organizational characteristics in system risk analysis for the chemical industry. [PLO 04], with MIRIAM1-ATHOS2, evaluated major risk management systems examining technical, human and organizational factors. [CHE 06] focused on the representation of accidental scenarios via the bowtie formalism to facilitate the organizational learning process and [MOH 09] proposed a means of carrying out probabilistic safety studies by taking organizational factors into account.

A recent method was proposed during an academic/industrial collaboration by EDF, INERIS (L'Institut National de l'Environnement Industriel et des Risques) and CRAN (Research Center for Automatic Control of Nancy) [LÉG 09]. This method is called integrated risk analysis. It focuses on a unified risk modeling. The model is multidisciplinary and generic. The generic property comes from a bowtie node for the technical parts, which serves as a basis from which to integrate knowledge about the human, organizational and environmental contexts through the barriers' efficiency that mitigate the influences of basic events on the consequences of accidental situations.

This modeling methodology has been applied to the assessment of risks in an industrial power plant and, more precisely, to the assessment of technical, human and organizational mitigation actions [LÉG 08a]. The unified model is structured based on the organizational level, the human actions level and their impacts (inhibition) on the propagation of causes and consequences into the bowtie model [LÉG 08a, LÉG 08,

LÉG 08b]. The model structure is given in Figure 2.13. The model obtained is used to estimate the occurrence probability of some risk scenarios and to assess the efficiency of barriers on risk reduction. In the risk management process [IEC 09], an engineer can use the model to identify the weaknesses of the socio-technical system and act accordingly to keep the risk criticality below an acceptable level. For the computation part, the model is based on a BN because it combines knowledge and observations to simulate scenarios and to identify weaknesses.

The reader can find an application of this BN-based modeling methodology on a chemical process in [LÉG 09]. In this application, 80 variables are considered. In [DUV 12], a critical system of a power plant is modeled with approximately 110 variables. This latter model is presented in Figure 2.14. The large number of variables in such a model means that only specialists are qualified to understand and perform simulations to characterize risks.

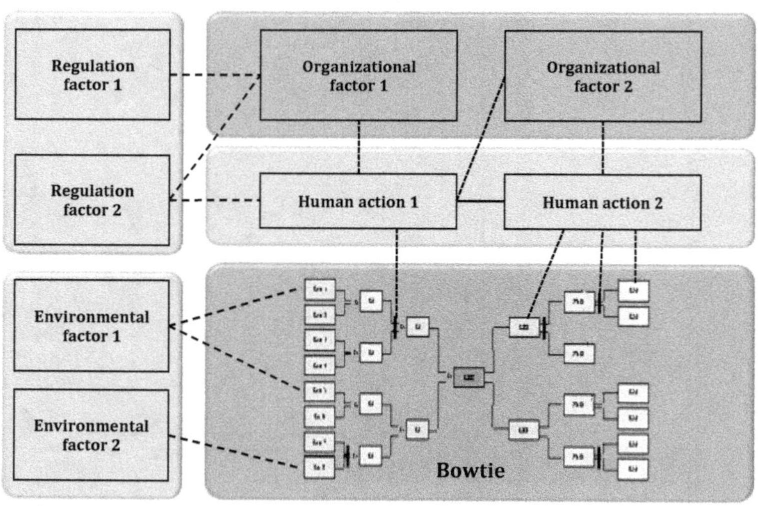

Figure 2.13. *Structuration in organizational level and action phases relating to a bowtie model*

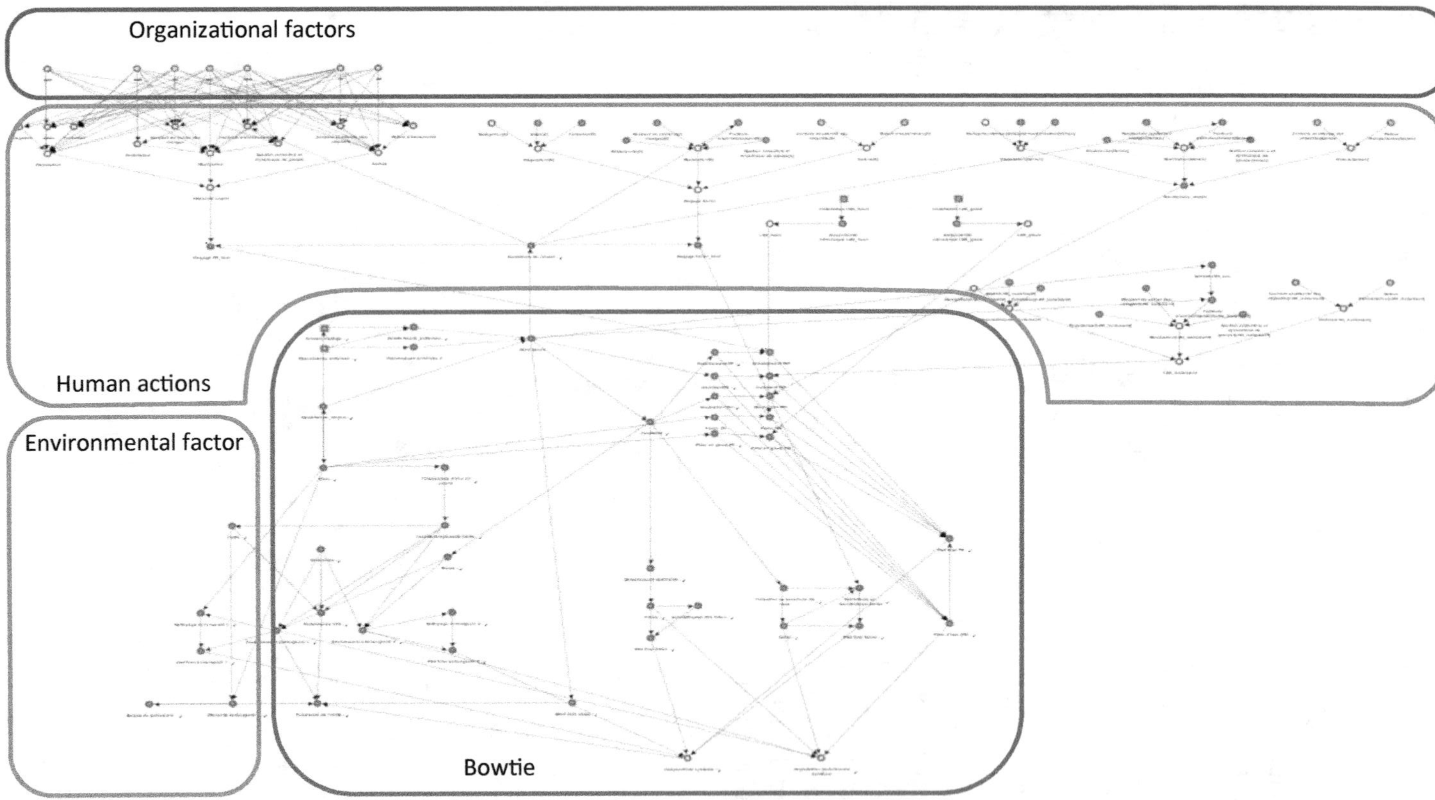

Figure 2.14. *RB unified model of the power plant risk*

2.6. Conclusion

This chapter illustrates how BN can solve the modeling problems of dependability and risk analysis of complex systems. This formalism works well with usual Boolean approaches such as FT or RBD. The construction of BN models in such cases follows the same guidelines and gives the same results.

The construction can be automatic by enumerating the cut-sets or tie-sets, whether minimal or not. Moreover, BN is well suited to modeling complex systems where dependencies between variables are not only deterministic.

Several BN structures are feasible for system dependability modeling. Model validation rests on the validation of the method used to build the model and confrontating the model with experiences by testing scenarios to validate the coherence of the model with well-known real cases.

Although BN provides a very compact model of large and complex problems, and makes it possible to handle hundreds of variables, some thousands of variables are needed for large industrial systems. In this case, the BN reach their limits. When the number of variables becomes too large, i.e. if the model cannot be supported in the memory of the computer that handles it, it is then necessary to use a more suitable modeling formalism.

3

Bayesian Network: Modeling Formalism of the Structure Function of Multi-State Systems

3.1. Introduction

In the case of multi-state systems, standard dependability methods, proposed in the literature are difficult to implement [LIS 03]. In this section, the methodology previously presented in the Boolean case is transposed to multi-state systems to prove that it is easy to obtain multi-state models with BN. Methods are presented for the construction of a model of multi-state systems. The methods are based on cut-sets, tie-sets or the principle of top-down analysis based on functional analysis. Section 3.2.3 explains the functional analysis-based method and explains how it provides an easy way to build an efficient model.

3.2. BN models in the multi-state case

The first step when modeling a multi-state model for dependability analysis is to define the set of variables x_i that represent the component states [SHU 10] as follows:

$$x_i = 0 \text{ if the component } i \text{ is in normal working state;}$$
$$x_i = 1 \ldots (l_i - 1) \text{ if the component } i \text{ is in a degraded state;} \quad [3.1]$$
$$x_i = l_i \ldots n_i \text{ if the component } i \text{ is in a failure state.}$$

with l_i being the first failure state, i.e. the component does not satisfy its functioning goals. States $1 \ldots (l_i - 1)$ are degraded functioning states, i.e. the component is not fully functional but it does not compromise the system mission. States $l_i \ldots n_i$ are several failure states of the component that can have different consequences on the system state.

The system state is also defined by a multi-state variable with respect to different functioning and malfunctioning scenarios. This variable is denoted as y and takes its values in the following states:

$$y = 0 \text{ corresponds to the well-functioning state;}$$
$$y = 1 \ldots (l - 1) \text{ correspond to degraded states;} \quad [3.2]$$
$$y = l \ldots n \text{ correspond to malfunctioning states.}$$

Regarding the complexity of scenarios in a multi-state system, it is difficult or impossible to model the system dysfunction using a FT or the functioning of the system by using a RBD. The analysis based on minimal cut-sets or minimal tie-sets remains efficient, and this method provides the definition of all the scenarios. The BN is an efficient modeling method by which to represent these scenarios. The purpose is to model the state of the system as a function of the components' states by the multi-state function ϕ. This function can be written as $y = \phi(x)$, where vector $x = (x_1, x_2, \ldots, x_r)$.

3.2.1. *BN model of multi-state systems from tie-sets*

The multi-state structure function is easily modeled by a BN. The variables mainly represent the components, the system and the scenarios. The second step is to structure the BN to efficiently link the component variables to the system variable, to encode the functioning and the failure scenarios of the system.

A first solution consists of enumerating all the minimal tie-sets or minimal cut-sets. By applying the same approach as in the binary case, a BN is defined to represent the conditional dependencies linking the system functioning or the failure states of the system with the minimal cut-sets or the minimal tie-sets. For the system shown in Figure 1.2, seven functioning scenarios exist: one is the perfect functioning state and the others are degraded functioning scenarios. In this modeling problem, the degraded states of the system are not modeled; therefore, the system state is defined only by two states: when the system is functioning $y = 0$; and $y = 1$ otherwise. The minimal tie-sets are defined from the following combination of components' states: 0 corresponds to Ok, 1 corresponds to Rc, and 2 corresponds to Ro as defined previously in Table 1.2 of Chapter 1:

$$\begin{aligned}
L_1 &= \{x_1 = 0, x_2 = 0\} \\
L_2 &= \{x_1 = 0, x_3 = 0\} \\
L_3 &= \{x_1 = 0, x_2 = 2\} \\
L_4 &= \{x_1 = 0, x_3 = 2\} \\
L_5 &= \{x_1 = 2, x_2 = 0, x_3 = 0\} \\
L_6 &= \{x_1 = 2, x_2 = 1, x_3 = 0\} \\
L_7 &= \{x_1 = 2, x_2 = 0, x_3 = 1\}
\end{aligned} \qquad [3.3]$$

Tie-set L_j is said to have occurred (been realized) if the components are in the states that define the tie-set. The occurrence of a tie-set has defined by $(L_j = 0)$. If at least one of the tie-sets is occurred, then the system is in the functioning state $y = 0$. The BN structure is obtained by linking each of the tie-sets L_j to the variables characterizing the components' states x_i involved in each tie-set (see Figure 3.1).

This approach automatically gives a BN, but the model obtained is not compact and, hence, is difficult to understand. To render the BN more compact, it is pertinent to combine nodes representing minimal tie-sets to the same variables such as $\{L_1, L_3\}$, $\{L_2, L_4\}$ and $\{L_5, L_6, L_7\}$ by creating complex variables and using the full capability of CPT based on multi-state logical combinations.

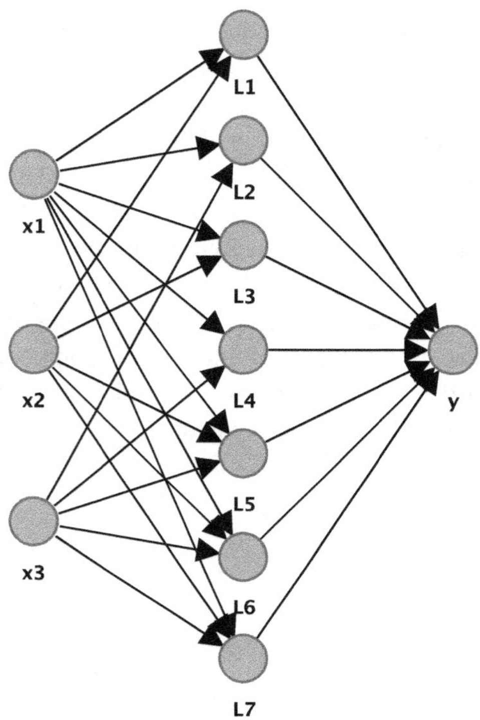

Figure 3.1. *BN structured by the minimal multi-state tie-sets*

For $\{L_1, L_3\}$, $L_{13} = L_1 \cup L_3$ is defined. $L_{13} = 0$ for the two scenarios: $L_1 = \{x_1 = 0, x_2 = 0\}$ and $L_3 = \{x_1 = 0, x_2 = 2\}$. In all other cases, $L_{13} = 1$ (see Table 3.1). For L_{24}, the CPT is defined in the same manner and is the same as for L_{13} (see Table 3.2). Finally, for L_{567}, the variable is in state 0 for each scenario: $L_5 = \{x_1 = 2, x_2 = 0, x_3 = 0\}$, $L_6 = \{x_1 = 2, x_2 = 1, x_3 = 0\}$ and $L_7 = \{x_1 = 2, x_2 = 0, x_3 = 1\}$ (see Table 3.3). This compact BN structure is given in Figure 3.2.

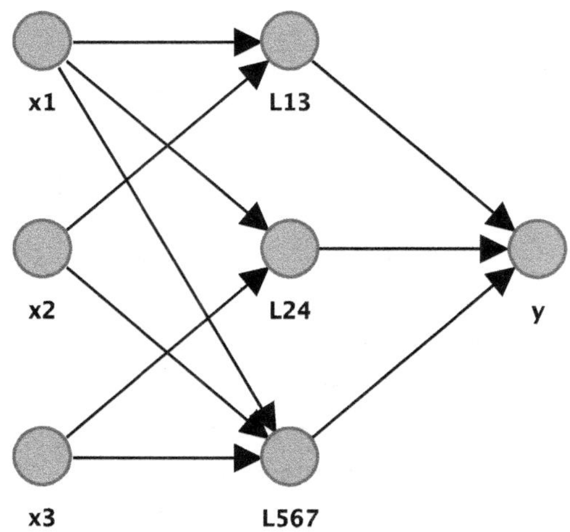

Figure 3.2. *Compact BN structured by minimal tie-sets for a multi-state system*

x_1	x_2	$P(L_{13}=0)$	$P(L_{13}=1)$
0	0	1	0
	1	0	1
	2	1	0
1	0	0	1
	1	0	1
	2	0	1
2	0	0	1
	1	0	1
	2	0	1

Table 3.1. *Multi-state L_{13} tie-set*

x_1	x_3	$P(L_{24}=0)$	$P(L_{24}=1)$
0	0	1	0
	1	0	1
	2	1	0
1	0	0	1
	1	0	1
	2	0	1
2	0	0	1
	1	0	1
	2	0	1

Table 3.2. *Multi-state L_{24} tie-set*

x_1	x_2	x_3	$P(L_{567}=0)$	$P(L_{567}=1)$
0	0	0	0	1
	0	1	0	1
	0	2	0	1
0	1	0	0	0
	1	1	0	0
	1	2	0	0
0	2	0	0	0
	2	1	0	0
	2	2	0	0
1	0	0	0	1
	0	1	0	1
	0	2	0	1
1	1	0	0	1
	1	1	0	1
	1	2	0	1
1	2	0	0	1
	2	1	0	1
	2	2	0	1
2	0	0	1	0
	0	1	1	0
	0	2	0	1
2	1	0	1	0
	1	1	0	1
	1	2	0	1
2	2	0	0	1
	2	1	0	1
	2	2	0	1

Table 3.3. *Multi-state L_{567} tie-set*

Through the inference mechanism, the probability distribution of y is computed and given by equation [1.3] and the probability distribution of the L_i tie-sets are given in Table 3.4.

State	$P(y)$	$P(L_{13})$	$P(L_{24})$	$P(L_{567})$
0	0.345721859	0.214953278	0.20012291	0.066539133
1	0.654278141	0.785046723	0.79987709	0.933460867

Table 3.4. *Results of the computation based on multi-state and tie-sets*

3.2.2. *BN model of multi-state systems from cut-sets*

The approach presented in section 3.2.1 can be applied to minimal cut-sets. For the system considered, the following four scenarios can be identified:

$$\begin{aligned} C_1 &= \{x_1 = 1\} \\ C_2 &= \{x_2 = 1, x_3 = 1\} \\ C_3 &= \{x_1 = 2, x_2 = 2\} \\ C_4 &= \{x_1 = 2, x_3 = 2\} \end{aligned} \quad [3.4]$$

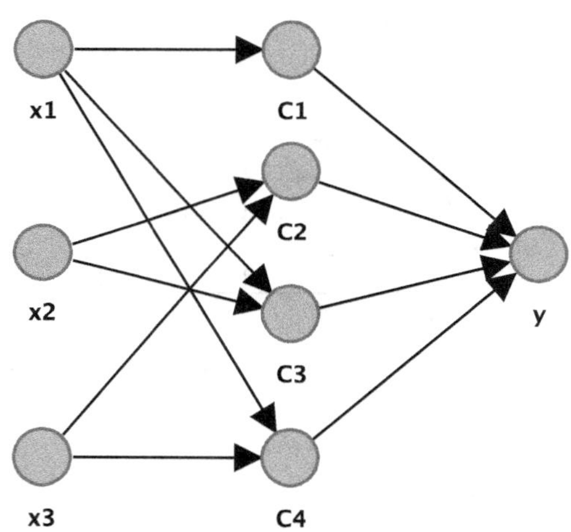

Figure 3.3. *BN based on the minimal cut-sets of a multi-state system*

The BN model structure is given in Figure 3.3. This structure is different from those built from the minimal tie-sets. This structure is compact and uses multi-states' modeling ability of BN. The CPT of the minimal C_1 to C_4 cut-sets are given in Tables 3.5, 3.6, 3.7 and 3.8.

x_1	$P(C_1 = 0)$	$P(C_1 = 1)$
0	0	1
1	1	0
2	0	1

Table 3.5. *Multi-state C_1 tie-set*

x_2	x_3	$P(C_2 = 0)$	$P(C_2 = 1)$
0	0	0	1
	1	0	1
	2	0	1
1	0	0	1
	1	1	0
	2	0	1
2	0	0	1
	1	0	1
	2	0	1

Table 3.6. *Multi-state C_2 tie-set*

x_1	x_2	$P(C_3 = 0)$	$P(C_3 = 1)$
0	0	0	1
	1	0	1
	2	0	1
1	0	0	1
	1	0	1
	2	0	1
2	0	0	1
	1	0	1
	2	1	0

Table 3.7. *Multi-state C_3 tie-set*

Through the inference mechanism, the probability distributions of any variable can be computed, y, and the minimal C_1 to C_4 cut-sets are given in Table 3.9. As the model is correct, the probability distribution for y is the same as computed previously (see equation [1.3]).

x_1	x_3	$P(C_4 = 0)$	$P(C_4 = 1)$
0	0	0	1
	1	0	1
	2	0	1
1	0	0	1
	1	0	1
	2	0	1
2	0	0	1
	1	0	1
	2	1	0

Table 3.8. *Multi-state C_4 tie-set*

State	$P(y)$	$P(C_1)$	$P(C_2)$	$P(C_3)$	$P(C_4)$
0	0.345721859	0.77218	0.88195459	0.780582261	0.776463366
1	0.654278141	0.22782	0.11804541	0.219417739	0.223536634

Table 3.9. *Results of the computation based on multi-state and cut-sets*

As shown in previous sections, a BN model can easily be built from the tie-sets or minimal cut-sets for any type of system: simple, complex, binary or multi-state. The models shown in Figures 3.1, 3.2 and 3.3 are equivalent, even if they do not exhibit the same variables. They are equivalent because they model the same joint probability distribution (y, x_1, x_2, x_3) with different factorizations.

As discussed previously, an automatic construction can be realized. Nevertheless, the models obtained are of large dimension and have no explicit structure. This structure of three layers with components, cut-sets or tie-sets and system missions is not of high interpretability, particularly for industrial systems.

3.2.3. *BN model of multi-state systems from functional and dysfunctional analysis*

A functional/dysfunctional approach can be used to build a BN without enumerating all functional/dysfunctional scenarios. A functional analysis like IDEF0 associated with a dysfunctional analysis, as proposed in [WEB 01, MUL 04, WEB 06, MED 13, MED 15], can serve to build a more readable structure. This approach is also well suited for multi-state systems.

Functional analysis of a system defines a model structure based on the functions achieved by the system. This analysis is interesting because it provides a model structure according to the levels of abstraction describing the functional architecture. Moreover, the system is not limited to the technical system, but can also include human or organizational levels [MED 11].

A function is achieved in a system if its environment provides the necessary input flows: operating conditions, operating supports, energy, orders, etc. Several input flows may contribute to the achievement of a function and the output flows represent the results of the function; thus, the pattern of a generic function is shown in Figure 3.4.

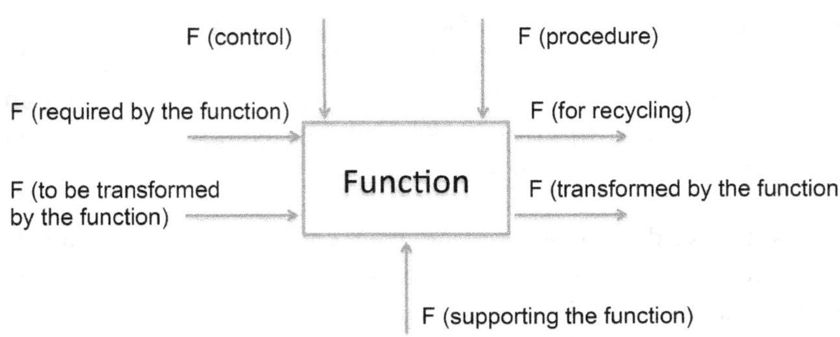

Figure 3.4. *Generic definition of a function and its flows*

Each part of the BN model corresponds to the functions defined by the functional analysis. Variables in the BN are defined for each input and output of each function. A generic BN pattern is defined in Figure 3.5 to model functions.

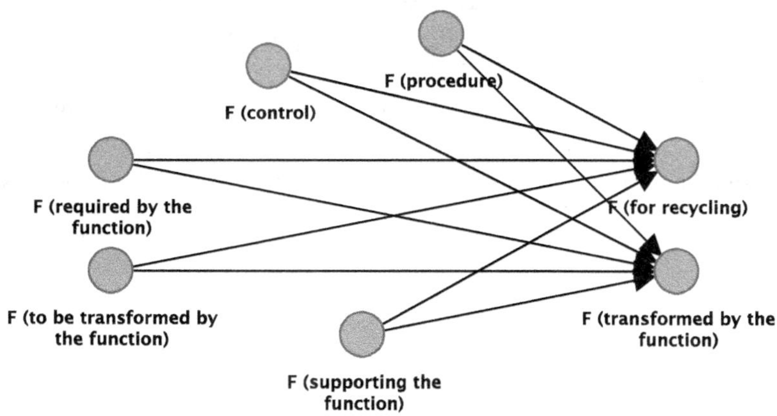

Figure 3.5. *Generic BN pattern of a function*

BN variables are multi-state and each of the conditional probability tables defines the dependency relationships between variables (not necessarily deterministic CPT). HAZOP and FMEA analysis specifies varying states as the failure modes (operating states, degraded states and failure modes), and these states represent the output flows of the function.

Input flows of the functions	Variables	Output flows
F (V1, V2, V3) ; F (water to transfer) ; F (control)	L	F (L)
F (V1) ; F (water to transfer) ; F (control)	L_1	F (L1)
F (L1) ; F (V2) ; F (control)	L_2	F (L2)
F (L1) ; F (V3) ; F (control)	L_3	F (L3)
F (V1, V2, V3) ; F (water to transfer) ; F (control)	I	F (I)
F (V1) ; F (water to transfer) ; F (control)	I_1	F (I1)
F (V2) ; F (water to transfer) ; F (control)	I_2	F (I2)
F (I2) ; F (V3) ; F (control)	I_3	F (I3)

Table 3.10. *Variables in the IDEF0 model representing the flow F(i)*

The modeling method is applied to the system shown in Figure 1.2. The functional analysis is given in Figures 3.6, 3.7, 3.8 and 3.9. From functional modeling, the variables x_1, x_2 and x_3 are defined to model the flows F (V1), F (V2) and F (V3), representing the states of the components of the system. The variable y is defined to model the flow F (water transferred) and represents the finality of the system (main function). A variable is defined for each function (Table 3.10) that models the output flow depending on the input flows of the functions.

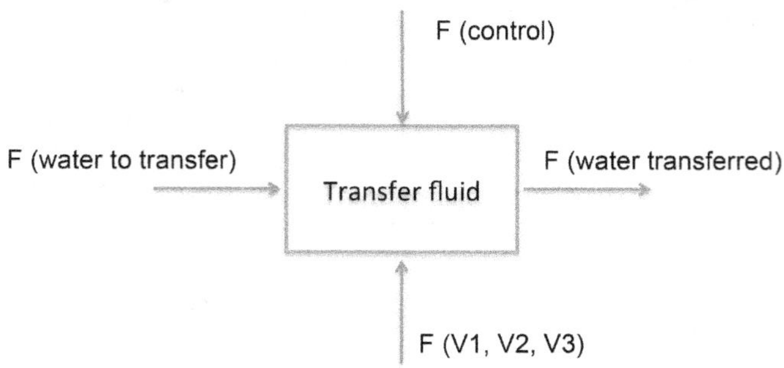

Figure 3.6. *Functional model of the system*

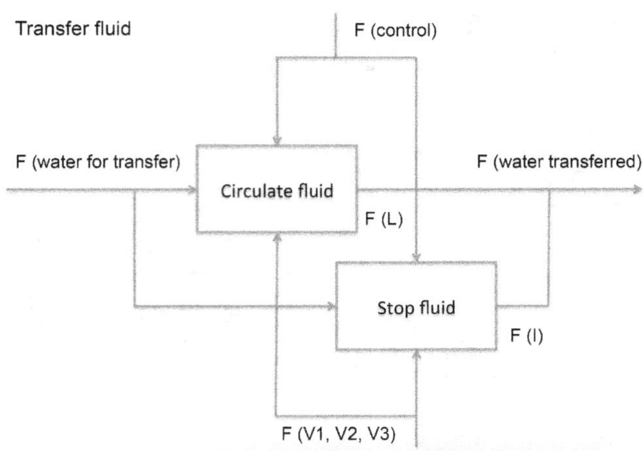

Figure 3.7. *Model of the function (transfer the fluid)*

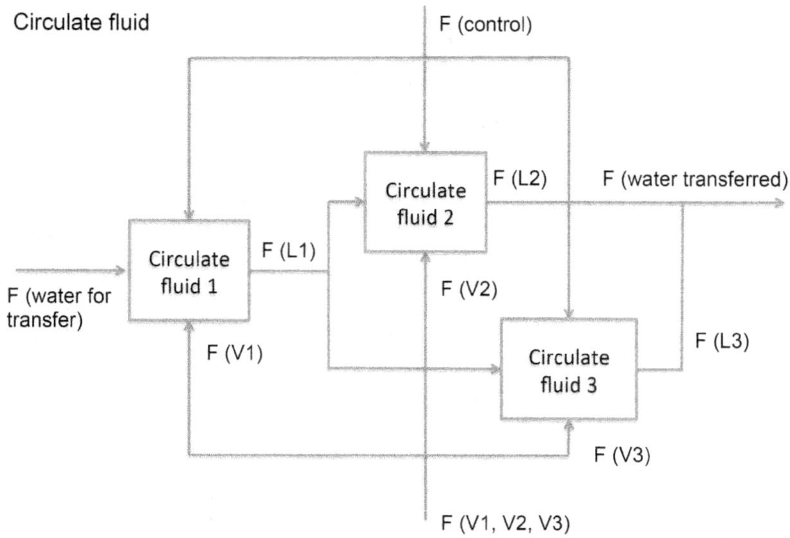

Figure 3.8. *Model of the function (circulate the fluid)*

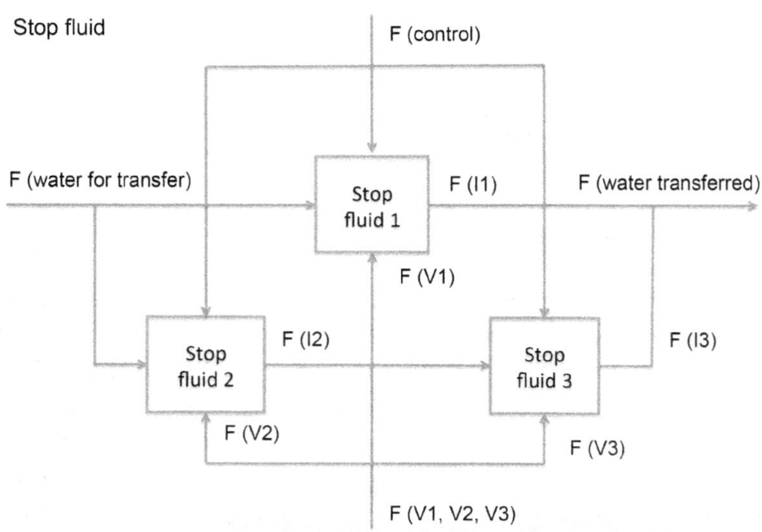

Figure 3.9. *Model of the function (stop the fluid)*

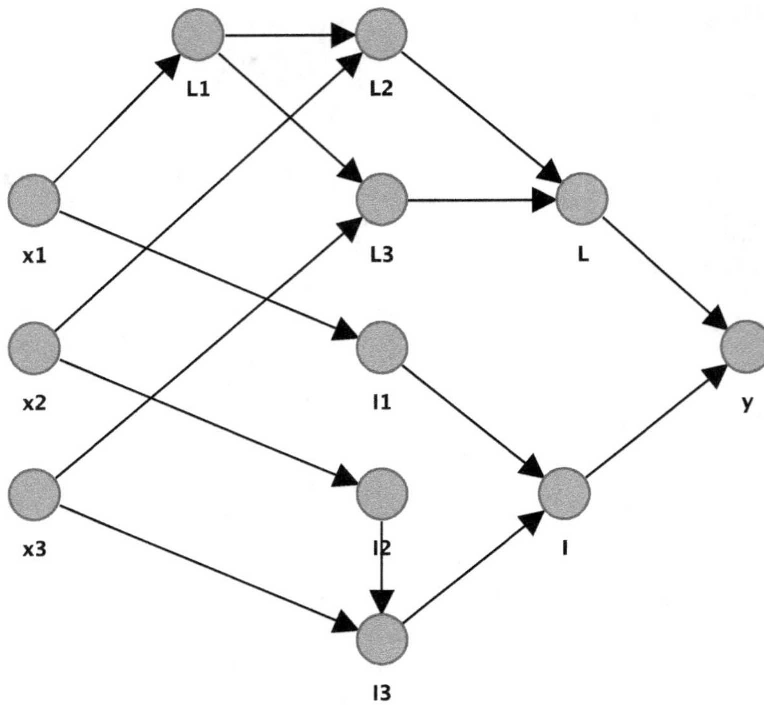

Figure 3.10. *BN model mapped from the functional model of the system*

The BN structure shown in Figure 3.10 is deduced from the flows defined in the functional analysis. For instance, functional analysis defines "circulate the fluid 2" function (Figure 3.8), and the output flow F (L2) is represented by the variable L_2 in the BN (Figure 3.10). Input flows of this function are F (L1) and F (V2) (for simplicity, F (control) is not modeled in this example); therefore, in the BN, the parents of variable L_2 are variables L_1 and x_2 that model the flows F (L1) and F (V2). The BN model is defined by connecting all the variables representing the input and output flows in the functional analysis. The resulting model is different from the models built from minimal cut-sets and minimal tie-sets; this time the model structure is based on a functional analysis of the system.

The inference in the BN model computes the probability y. The results are equal to those computed from the cut-sets, the tie-sets or from the initial joint law. The probabilities of the variables L and I modeling "circulate the fluid" and "stop the fluid" functions of the system, are presented in Table 3.11. The variables L_1, L_2, L_3, I_1, I_2 and I_3 are presented in Tables 3.12 and 3.13.

State	$P(y)$	$P(L)$	$P(I)$
0	0.345721859	0.681027695	0.664694164
1	0.654278141	0.318972305	0.335305836

Table 3.11. *Results of the computation based on the IDEF0 model*

State	$P(L_1)$	$P(L_2)$	$P(L_3)$
0	0.77218	0.524348829	0.488172196
1	0.22782	0.475651171	0.511827804

Table 3.12. *Results of the computation based on the IDEF0 model L_i variables*

State	$P(I_1)$	$P(I_2)$	$P(I_3)$
0	0.54437	0.51843	0.264083058
1	0.45563	0.48157	0.735916942

Table 3.13. *Results of the computation based on the IDEF0 model I_i variables*

The proposed approach to building the BN structure of the industrial system is a generalization of the other modeling methods to the multi-state case. This method is suitable for Boolean models as well as multi-state models. The method is based on the hierarchical description of the system functions because the chosen functional analysis provides the hierarchical description of the system functions; nevertheless, this principle may be transposed to other functional analyses to avoid this hypothesis.

3.3. Non-deterministic CPT

For a multi-state system, the relation between the variable y and its parents x_i can be non-deterministic, as mentioned in section 2.4. If the conditional probabilities defining y are in the $0, 1$ interval then the CPT is deterministic. However, if these conditional probabilities are in the $]0, 1[$ interval, then the CPT is non-deterministic. As in the binary case, this kind of CPT means that the expert is not completely sure that the occurrence of x_i leads to the occurrence of y. Non-deterministic CPT is encountered for several reasons: the relation between x_i and y is naturally non-determinist or some parents (x_i) are missing from the model. The inability of an expert to define the relation between x_i and y with complete certainty is translated into a non-deterministic CPT.

Let us illustrate this concept using an industrial example. The Omega-20 methodology is dedicated to modeling human safety barrier (HSB) performance. As mentioned in [MIC 09], the assessment of the performance aims to determine the level of confidence in the barrier. The probability of efficiency of the HSB corresponds to a risk reduction factor of the critical event propagation. The HSB is Efficient or Not Efficient; if the HSB is Efficient, the propagation of the critical event is reduced by 100%, and the occurrence of this accident becomes equal to 0; if the HSB is Not Efficient, the critical event propagation is not reduced, and the occurence of the accident is not affected by the HSB.

Moreover, the HSB is based on three steps: detection, diagnosis and action. Each of these steps has a performance classified as follows: 0, 1 or 2. The barrier acts to inhibit the critical event. As shown in Figure 3.11, the HSBs reduce the probability that an event x_A propagates its effect to the output y. If the detection is inefficient (confidence level 0), the diagnostic is of low quality and the action has a high stress level, and the event can propagate fully. If all these steps are at their best level, the efficiency probability is equal to 1 and the event x_A cannot be propagated. When one of the steps is at level 1, it divides the probability of the critical event propagation by 10, and by

100 at level 2. Then, if detection and diagnostic are at level 1, then the HSB efficiency has a probability equal to 0.01. Therefore, the efficiency probability takes values from 0.000001 to 1, as defined in Table 3.14.

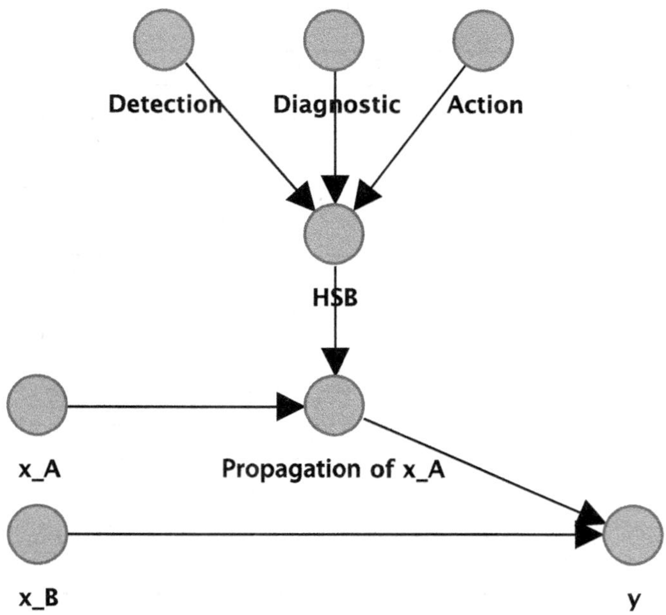

Figure 3.11. *BN model of a human safety barrier (HSB)*

This methodology is used in the protection and prevention phases in a bowtie model. For instance, it has been used in [LÉG 09, DUV 12] for the integrated risk analysis methodology (Figure 2.14) applied to a chemical process. According to OMEGA20 methodology, it can easily be introduced in many technical systems modeled by a bowtie.

3.4. Industrial applications

The advantages of this modeling approach are particularly interesting when large systems are modeled. This section discusses an application of the modeling method to an industrial case to provide a model for decision-making in maintenance strategy evaluation. The maintenance process is fundamental to improving the availability and productivity of industrial systems. To control these performances, maintenance managers need to be able to choose a maintenance strategy and adequate resources to perform this strategy.

$Detection$	$Diagnostic$	$Action$	$P(HSB = Efficient)$	$P(HSB = NotEfficient)$
0	0	0	1	0
0	0	1	0.1	0.9
0	0	2	0.01	0.99
0	1	0	0.1	0.9
0	1	1	0.01	0.99
0	1	2	0.001	0.999
0	2	0	0.01	0.99
0	2	1	0.001	0.999
0	2	2	0.0001	0.9999
1	0	0	0.1	0.9
1	0	1	0.01	0.99
1	0	2	0.001	0.999
1	1	0	0.01	0.99
1	1	1	0.001	0.999
1	1	2	0.0001	0.9999
1	2	0	0.001	0.999
1	2	1	0.0001	0.9999
1	2	2	0.00001	0.99999
2	0	0	0.01	0.99
2	0	1	0.001	0.999
2	0	2	0.0001	0.9999
2	1	0	0.001	0.999
2	1	1	0.0001	0.9999
2	1	2	0.00001	0.99999
2	2	0	0.0001	0.9999
2	2	1	0.00001	0.99999
2	2	2	0.000001	0.999999

Table 3.14. *CPT of HSB efficiency*

A decision-making application in maintenance is proposed by Medina-Oliva [MED 11]. The author formalizes a methodology to develop a model to evaluate and compare different maintenance strategies. The model required merges many complementary views of the system: a functional view of the system, a dysfunctional view of the system, an organization view of the maintenance department and the technical maintenance team, and the effectiveness of its action policy according to the logistics.

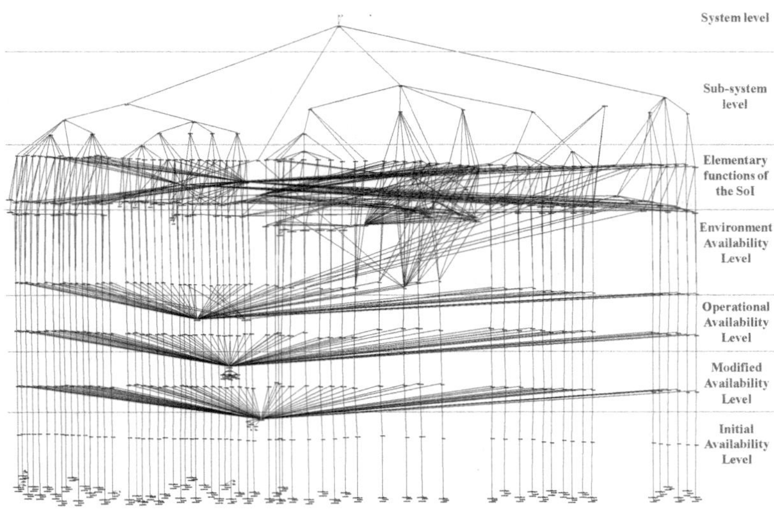

Figure 3.12. *BN model of the interaction between maintenance and performance of a food system*

It is impossible to formalize such a model as a monolithic set of interconnected variables. The model structure is based on fusion of the technical description of the functional view as described in the previous section and the integration of the human and organizational layer presented in Léges *et al.'s* PhD thesis [LÉG 08a, LÉG 09]. The methodology lies in the unification of different types of knowledge required for the construction of this model [MED 13, MED 15]. In this application, the BN reaches its limit; therefore, a probabilistic relational model (PRM) language is used to define the BN model and a specific inference algorithm is used to compute the probabilities in this very large model.

The model structure is built by the instantiation of several generic and modular patterns. The patterns are dedicated to modeling the decision variables for maintenance and the variables of the industrial system [MED 11]. This method facilitates the structuring and validation of the model.

Medina–Oliva's PhD these [MED 11, MED 13] describe an application of the method to modeling the interaction between maintenance and performance of a food system. The model is shown in Figure 3.12 (with 700 variables). The computation is only performed on a specific part of the model built partially to answer queries. The global model is never represented entirely by the inference algorithm. This modeling methodology results from the ANR SKOOB [ANR 11] project, and the modeling method is applied in partnership with SOREDAB on the maintenance of a food system [MED 13, MED 15].

3.5. Conclusion

This chapter illustrates how BN can solve the modeling problems of dependability and risk analysis of complex systems. This formalism is well suited to modeling complex multi-state systems where dependencies between variables are not only deterministic. Modeling with BN is an efficient and flexible solution for analysis of simple to complex systems and from binary to multi-state systems.

The modeling methodology based on functional and dysfunctional analysis proposed in this book coupled with a PRM is a very efficient modeling method for large systems, as confirmed by the applications of the ANR SKOOB project [MED 13, MED 15, ANR 11]. Inference algorithms based on PRM are still being developed [SOM 10, KN 10, GON 11]; nevertheless, a PRM modeling platform is provided by Bayesia-BRICKS (Bayesian Representation and Inference for Complex Knowledge Structuring). The PRM provides a framework for knowledge capitalization by creating generic classes and by instantiation of the model to particular systems. This modeling formalism is the future of probabilistic graphical models.

The models presented in this chapter are based on static representations, they do not take into account the temporal dimension. In dependability, it is important to model the temporal dimension to incorporate the impact of the environment, the system operating conditions, the aging and the degradation of components. Such modeling allows evaluation of the behavior of the probabilities of events occurring over time in order to predict future situations. The objective may be, for example, to anticipate the system degradations. This temporal dimension is presented in the next chapter.

PART 2

Dynamic Bayesian Networks

4

Dynamic Bayesian Networks: Integrating Environmental and Operating Constraints in Reliability Computation

4.1. Introduction

To take into account the uncertainty with the temporal aspect, system behavior is modeled by a random variable, which takes its values in finite states corresponding to the system states. The state space method is well known in dependability literature [COR 75, VIL 88, ANS 94, COC 97, AVE 99, GER 00] and also in industrial standards [IEC 06]. The models obtained estimate the failure probabilities of systems throughout their lifetime.

This method gives a graphical representation [VIL 88, p. 303; COC 97, p. 282] whose complexity depends on the hypotheses made that correspond to the real stochastic process. However, the complexity increases tremendously when the number of components increases. Indeed, the state space describing the system is built from the Cartesian product of the component states.

To reduce this problem, state aggregation techniques are proposed [COC 97, p. 282]. The Markov chains (MCs) obtained are usually compact. For instance, in [WEB 06, WEB 08, POU 09, WEB 12a] it was applied to wind turbine modeling and gives models with fewer states than the initial model. A new modeling approach has to satisfy such a compactness target.

Dynamic fault trees (DFTs) [DUG 92, MES 02] are one solution allowing us to describe a MC of great dimension. The DFT modeling is based on a graphical description language of the component combination (for instance, a passive redundancy structure, i.e. "spare gate"). The MC is generated by compiling the DFT model.

It is now necessary to model increasingly complex relations to estimate the dynamic behavior of systems. The models aim to estimate the state probability distributions of the system, taking into account the component age, the maintenance and control actions, and the context evolution. A holistic modeling of systems and their interaction with the context leads us to take into account a huge number of modeling states. This number of states makes the MC modeling very hard to realize because of the combinatory explosion of states. The model, even if it can be made by automaton combination, is very hard to read, understand, interpret and maintain [DE 92]. DFTs also have their limitations because they are based on a graphical language with the binary state hypothesis, and thus cannot conveniently model multi-state systems.

In this situation, BN are of interest because they extend the usual modeling capacity of graphical modeling to handle complex multi-state systems. Dynamic Bayesian networks (DBN) are known to be able to formalize stochastic processes in a compact form [MUR 02]. Thus, they solve the main limits of usual modeling methods. The first research works on the application of DBN to the reliability and system availability are given in [WEL 00]. In 2002, [WEB 02, WEB 03] applied DBN to the availability analysis of a system and showed the ability of a DBN to model a multi-state MC in a compact way. With this compactness property, a better model can be computed by reducing the modeling difficulty, to address more complex systems.

The goal of this section is to show the application of DBN in the dependability area and to explain the extension they allow according to the MC modeling and also their limits.

4.2. Component modeled by a DBN

A DBN can take into account the temporal dimension of the system states' evolution along their lifetime by factorizing and discretizing the state space of each independent random variable at each time instant. A stochastic process is represented at time k by a variable x_i^k with a finite number of states $(h_1^x, \dots h_n^x)$. The state of variables with the same value of k constitutes the time slice k [HUN 99, BOU 99].

A DBN can model the evolution of discrete random variables by defining the conditional dependence of a time slice $k + 1$, given the states of the random variables at the previous time slice k. The definition of the dependence linking the variables at different time slices can model various complex stochastic processes. This time-based stochastic process is modeled by a CPT. Figure 4.1 shows a particular case where a variable, x_i^k, is defined conditionally to itself at the previous time slide x_i^{k-1}. This is the Markovian case.

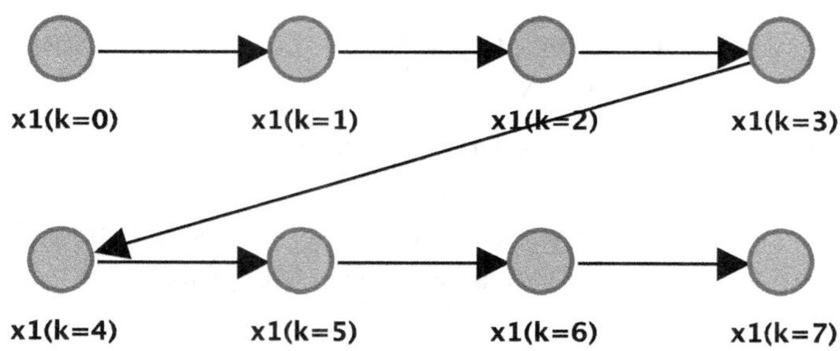

Figure 4.1. *DBN model developed over eight time slices*

From an observed situation at any time k or from the initial conditions with $k = 0$, the inference mechanism in the DBN allows us to compute the state probability distribution of all variables for each time slice. To compute this, it is necessary to memorize the state probability distribution of all the variables in all time slices. The solution consists of developing the time slices for the entire desired time horizon, i.e. to duplicate all the variables for each time period. However, the BN size increases proportionally to the computing horizon [KJA 95]. This solution is not convenient for system dependability analysis because the process should be studied for a large time horizon. It conducts to a combinatory explosion of variables that cannot be handled by current inference mechanisms.

4.2.1. *DBN model of a MC*

In the case of Markovian processes, the Markov property is used to simplify the inference mechanism. For instance, in a Markovian process, the CPT is time invariant. The inference can be realized iteratively without explicitly defining a variable for each time slice. The DBN model is then compact and only two successive time slices are modeled, as shown in Figure 4.2. A DBN with two time slices noted 2-TBN [BOY 98] allows us to define all the necessary parameters to model the MC. The first slice contains the variables at the current time step k, while the second allows us to compute the distribution of variables at the time step $(k + 1)$ by inference. A variable $x_i^{(k+1)}$ is defined conditionally to its states in the current time step $x_i^{(k)}$. The CPT, $x_i^{(k+1)|x_i^{(k)}}$, is constant whatever the value of k (Table 4.1). This CPT is defined from the transition probability matrix between the states of the MC. With this model, the future states at $(k + 1)$ are conditionally independent of the past given the present states at time (k). The CPT clearly shows a MC [KJA 95].

Figure 4.2. *DBN of a MC*

After the first inference, the distribution $P\left(x_i^{(k+1)}\right)$ is injected as the *a priori* distribution for $x_i^{(k)}$. The next inference allows us to compute the distribution for the next time step. An exact inference computes the probability distribution of the random variable for the time step $k + 1$, from the distribution at time step k. The probability distribution for the next time steps $k + 2, k + 3, \ldots$ are computed by successive inferences [WEL 00]. For a time horizon of size h, h inferences are necessary. This computing method is equivalent to the Chapman–Kolmogorov equation.

$x_i^{(k)}$	$P\left(x_i^{(k+1)} = 0\right)$	$\ldots$	$P\left(x_i^{(k+1)} = n_i\right)$
0	$P\left(\left(x_i^{(k+1)} = 0\right) \vert x_i^{(k)} = 0\right)$		$P\left(\left(x_i^{(k+1)} = n_i\right) x_i^{(k)} = 0\right)$
$\ldots$			
n_i	$P\left(\left(x_i^{(k+1)} = 0\right) x_i^{(k)} = n_i\right)$		$P\left(\left(x_i^{(k+1)} = n_i\right) x_i^{(k)} = n_i\right)$

Table 4.1. *CPT defining the transition probability matrix of a MC*

4.2.2. *DBN model of non-homogeneous MC*

The extension to non-homogeneous MC is possible by introducing time-indexed CPT. By working with the Bayesia company (http://www.bayesia.com/) this possibility has been introduced in the BayesiaLab software. The parameters defined in the CPT can be indexed to an exogenous variable k that represents time.

Here we illustrate the concept. Let us consider valves with three states: a normal functioning state and two failure states, i.e. a remained closed state $\{1\}$ and a remained open state $\{2\}$. In the case of varying parameters, the principle is illustrated by combining two Weibull laws for the valve x_1. The failure rates are time varying and defined according to Weibull laws with the following parameters:

– for the transition to the remained closed state $\{1\}$, the failure rate is defined as follows:

$$\lambda_{11} = \frac{\beta \times k^{(\beta-1)}}{\alpha^{\beta}} \text{ with } \beta = 3 \text{ and } \alpha = 500 \qquad [4.1]$$

– for the remained open state, the failure rate is defined as follows:

$$\lambda_{12} = \frac{\beta \times k^{(\beta-1)}}{\alpha^{\beta}} \text{ with } \beta = 2.5 \text{ and } \alpha = 700 \qquad [4.2]$$

The DBN model of the valve x_1 is shown in Figure 4.3. The probability distribution on the valve states is computed over 1,000 hours with 1,000 iterations, i.e. with a time step of 1 hour, as shown in Figure 4.4.

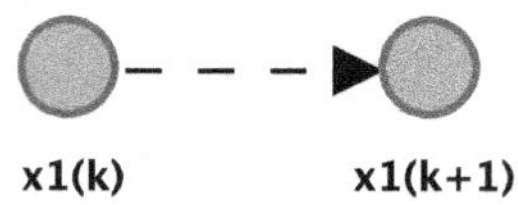


```
P(x1(k)=0|t=0) = 100
P(x1(k)=1|t=0) = 0
P(x1(k)=2|t=0) = 0

P(x1(k+1)|x1(k),t) =
 If(x1(k+1)==0 & x1(k)==0, 100-
((3*t^2/500^3)+(2.5*t^1.5/700^2.5))*100,
  If(x1(k+1)==1 & x1(k)==0, (3*t^2/500^3)*100,
   If(x1(k+1)==2 & x1(k)==0, (2.5*t^1.5/700^2.5)*100,
    If(x1(k+1)==0 & x1(k)==1, 0,
     If(x1(k+1)==1 & x1(k)==1, 100,
      If(x1(k+1)==2 & x1(k)==1, 0,
       If(x1(k+1)==0 & x1(k)==2, 0,
        If(x1(k+1)==1 & x1(k)==2, 0, 100))))))))
```

Figure 4.3. *DBN of a non-homogeneous MC*

4.2.3. *Stochastic process with exogenous constraint*

As shown in [WEB 04], a hidden Markov model (HMM) [RAB 89] can represent the degradation of components. The modeling of component degradation by HMM has also been used in [MOG 12, ROB 13, LE 14].

Time is usually considered as a conditional factor in component reliability, as shown in the previous section. It can be insufficient [SIN 95]. The conditions of use and the environmental context (like humidity, temperature, etc.) can alter the component reliability. All factors that alter component reliability are called co-variables or exogenous variables [BAG 01]. As described in [COX 55], the component reliability can be modeled precisely by taking into account the effects of exogenous variables.

In [WEB 04], several models of MC are defined according to the operational context of the component. A Markov switching model (MSM) is introduced to model the switching from one MC to another subsequent to the state variation of the exogenous variables. These models are also considered as conditional MC where the transitions are conditional to exogenous variables.

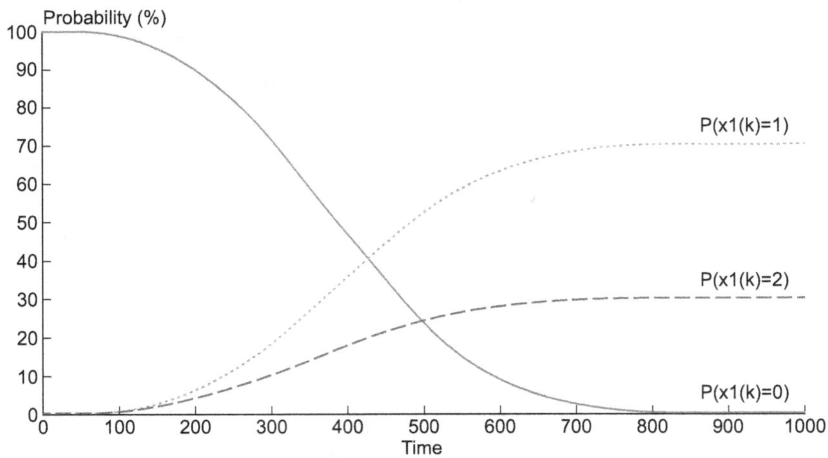

Figure 4.4. *Inference in the DBN of a non-homogeneous MC*

The MSM models are non-stationary because of the fast modifications of parameter values [BEN 99, p. 147]. A MSM represents the conditional distribution $P\left(x_i^{(k)}, u_i^{(k)}\right)$ given the input state sequence $\left(u_i^{(0)}, u_i^{(1)}, \ldots u_i^{(k)}\right)$, where $u_i^{(k)}$ represents the state of the exogenous variable. The simulation of the MSM is based on discontinuous changes of parameters associated with each modification of the exogenous variable state. It is very hard to obtain an analytical solution as with MC, and it is quite simple to use a simulation.

The modeling solution by a DBN is really simple [WEB 04]. One or several exogenous variables modeling the constraints or the operational conditions are added as new discrete variables $u_i^{(k)}$ in the time slice k. The CPT that defines the transition between two consecutive time steps, $x_i^{(k+1)}|x_i^{(k)}$, is defined conditional on $u_i^{(k)}$, as shown in Figure 4.5 for one variable.

Moreover, if the variable $x_i^{(k)}$ is not observed and only a sequence of the exogenous variable $u_i^{(k)}$ is observed, the stochastic process model of the component and its environment is formalized as a HMM conditioned by an input sequence $u_i(k)$. This model is called an input–output hidden Markov model (IOHMM). The exogenous variable $u_i^{(k)}$ is an input that models the constraints or the variations of the component environment. The impact of the hidden process on the failure mode is defined by the output $z_i^{(k)}$. The variables $u_i^{(k)}$ and $z_i^{(k)}$ induce the behavior of the hidden process that describes the component degradation but remains unobservable $x_i^{(k)}$ [BEN 06]. The formal model of the IOHMM is well suited to model complex stochastic processes such as $P\left(z_i^{(k)}, x_i^{(k)}, u_i^{(k)}\right)$ [BEN 99, p. 145].

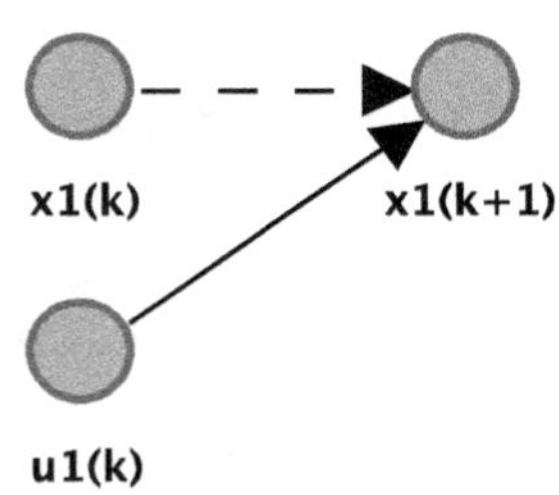

Figure 4.5. *DBN model of a MSM*

As proposed in [BEN 06], the DBN model of an IOHMM can easily be applied to the components' reliability (Figure 4.6). In this model, $u_i^{(k)}$ has three states:

– '0', the component is not used;

– '1', the component is used in normal conditions;

– '2', the component is used outside its nominal conditions.

A simulation of the component state evolution $x_i^{(k+1)}$ and of the function realized by the component $z_i^{(k+1)}$ is given for a sequence of $u_i^{(k)}$ defined from $k = 0$ to $k = 600$. The reliability of the component is defined as $y(k) = z_i^{(k)}$ (Figure 4.7).

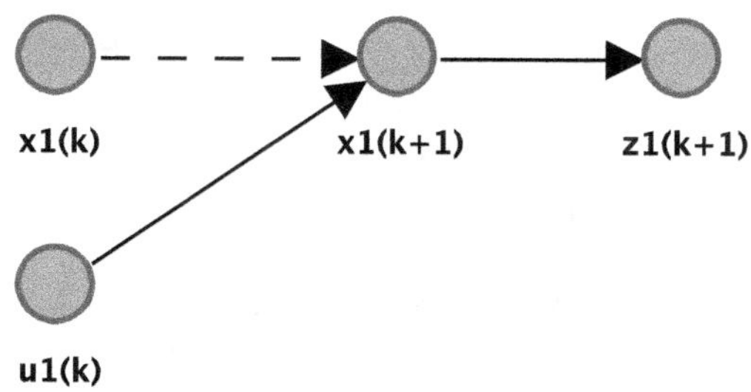

Figure 4.6. *DBN model of an IOHMM*

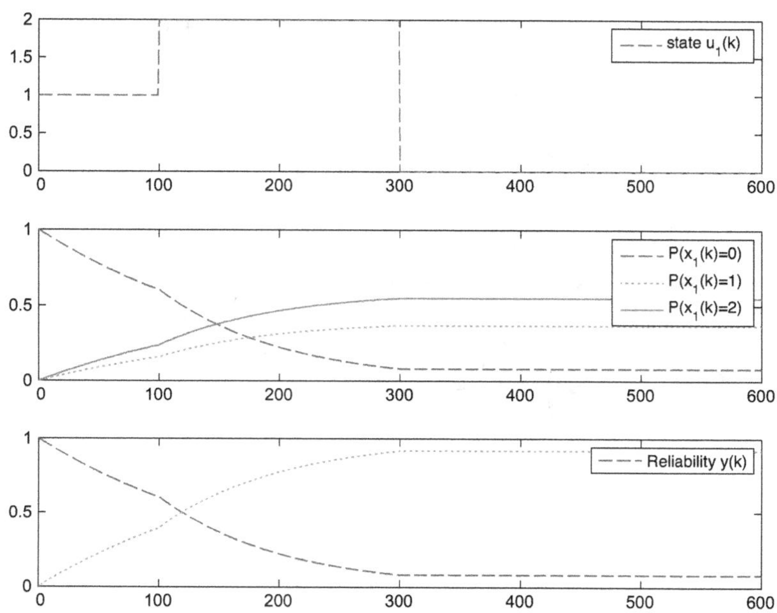

Figure 4.7. *Inference in the DBN model of the IOHMM*

4.3. Model of a dynamic multi-state system

A DBN is particularly interesting when dealing with several components in a system. The DBN presented in section 4.2 allows us to represent several multi-state stochastic processes in a system model. A multi-state model, as presented in Chapter 3, can easily combine the models of dynamic multi-state components as presented in section 4.2

to give a whole model of the dynamic multi-state system. The computation in a DBN with several stochastic processes is solved by different inference mechanisms well suited to this problem and to the conditions of use of the models.

The exact inference algorithms are based on a junction tree [JEN 96]. This mechanism is applied to unroll up models. If all time slices are defined in the model, then the usual inference algorithm can be used to compute the exact results (Figure 4.8) but with high computation costs. The models can be of high complexity with much dependence between the components (Figure 4.9) and thus are not practical for such a modeling step. Moreover, they are not adapted for a large time horizon.

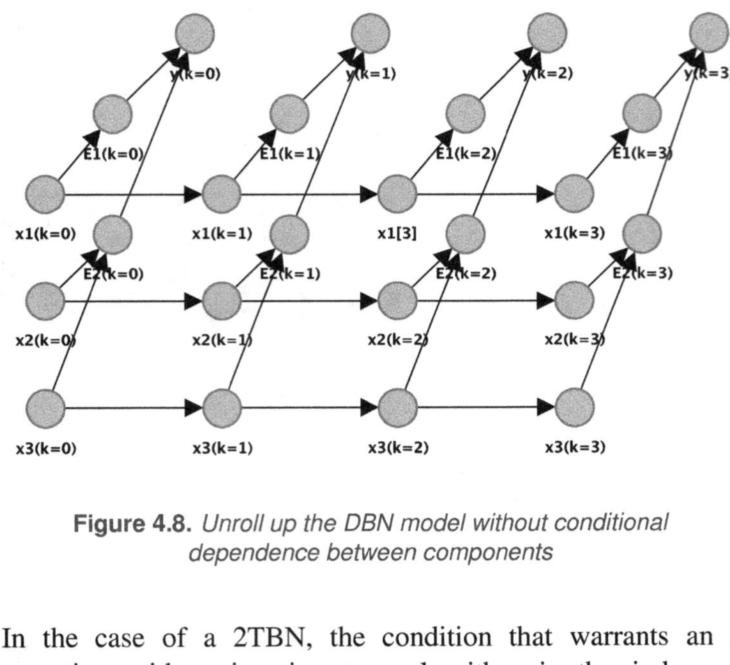

Figure 4.8. *Unroll up the DBN model without conditional dependence between components*

In the case of a 2TBN, the condition that warrants an exact computation with a junction tree algorithm is the independence between the root stochastic processes. This condition is verified when all the components are independent as in Figure 4.10. The marginal distribution is easily computed in a 2TBN by using the exact inference algorithm. The state probability distribution is computed for each multi-state component, as shown in Figure 4.11.

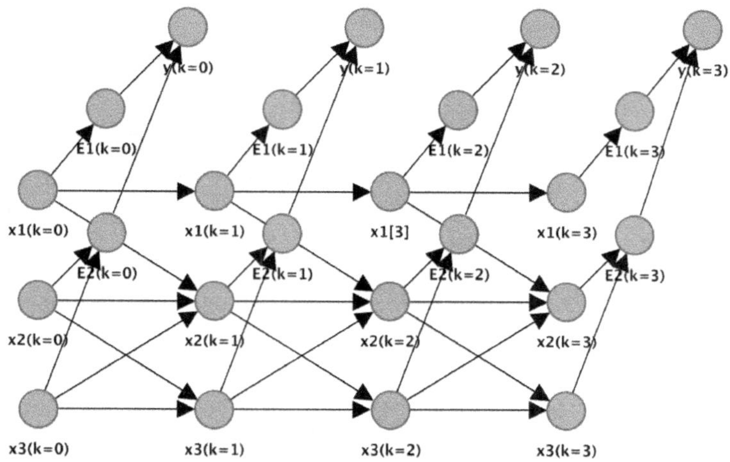

Figure 4.9. *Unroll up the DBN model with conditional dependence between components*

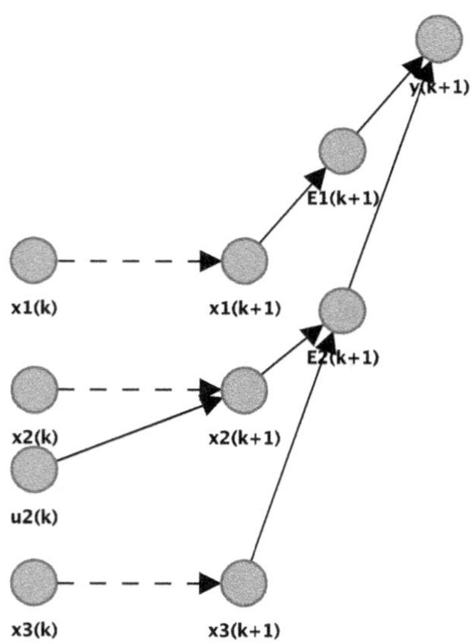

Figure 4.10. *2TBN of a multi-state system*

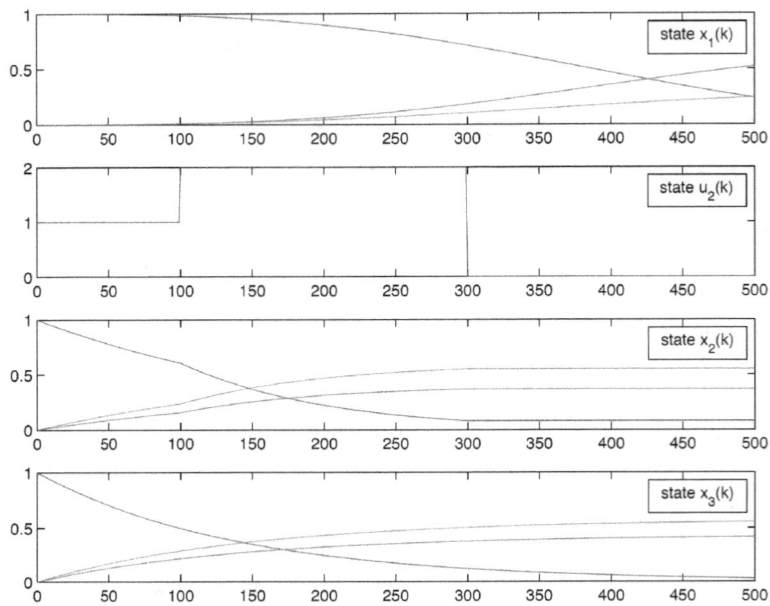

Figure 4.11. *Inference of a 2TBN multi-state model and state probability distribution*

For component x_1, the process is non-homogeneous:

– failure 1 (valve remained closed) has an occurrence probability given by equation [4.1];

– failure 2 (valve remained open) has an occurrence probability given by equation [4.2].

Component x_2 is conditionally defined to three functioning modes by the variable $u_2^{(k)}$:

– $u_2^{(k)} = 0$, the component is not used: $\lambda_{21} = 0$ and $\lambda_{22} = 0$;

– $u_2^{(k)} = 1$, the component is used in nominal conditions: $\lambda_{21} = 2.10^{-3}$ and $\lambda_{22} = 3.10^{-3}$;

– $u_2^{(k)} = 2$, the component is used outside its nominal conditions: $\lambda_{21} = 4.10^{-3}$ and $\lambda_{22} = 6.10^{-3}$.

Component x_3 is modeled by a homogeneous MC with the following transition probabilities $\lambda_{31} = 3.10^{-3}$ and $\lambda_{32} = 4.10^{-3}$.

Then, the independent components are modeled in the DBN by a non-homogeneous MC, an IOHMM, and a homogeneous MC. The DBN models the multi-state system composed of the independent components in a compact form. The models of the processes are shown in Figure 4.10. The time slide $k + 1$ computes the state propagation from the components to the system $y(k)$, to compute its reliability (Figure 4.12).

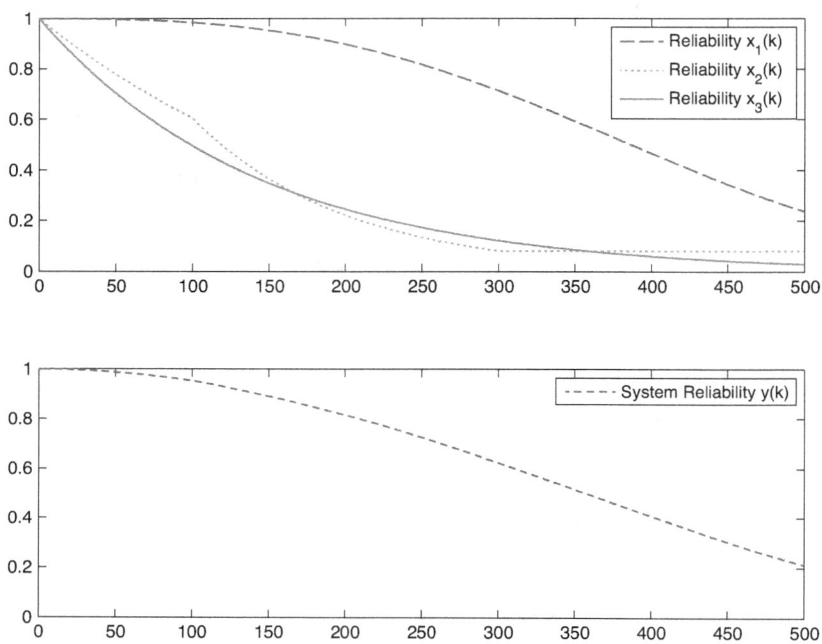

Figure 4.12. *Multi-state system and components' reliability*

4.4. Discussion on dependent processes

The components of systems are not always independant. To decrease the model complexity in the case of dependent processes, it is possible to mix the dependent components in only one stochastic process that is combined with other independent processes by a multi-state BN, as shown previously. According to this method, the

DBN models only independent processes. The whole structure of the global system is then simplified, but the number of states of some variables increases.

Nevertheless if some dependence exists between stochastic processes, as in the roll up of DBN shown in Figure 4.9, it is necessary to use a specific inference algorithm that computes the joint distribution at each time step with significant computing and memory costs. The approximate inference algorithm proposed by [BOY 98, KOL 99] or particular filtering [KOL 00] can estimate the marginal distribution with a bounded error, which is often sufficient for dependability purposes.

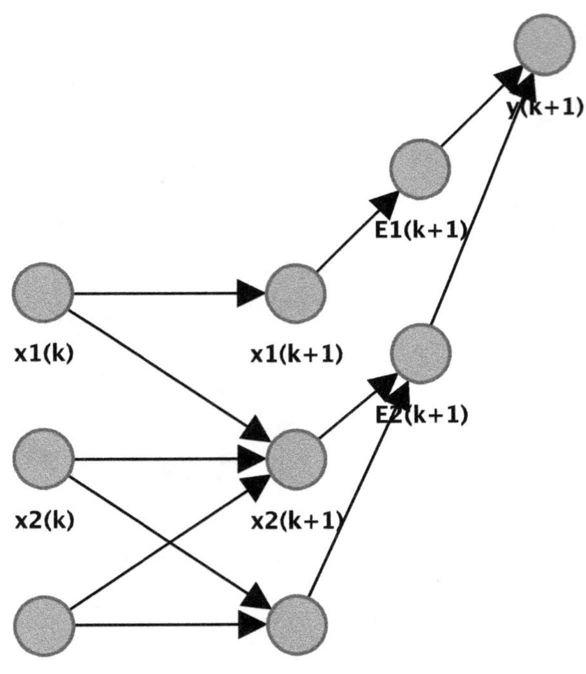

Figure 4.13. *2TBN model of a multi-state system with largely dependent processes*

Unfortunately, another phenomenon introduces difficulties in computing the marginal distribution even in the case of the independent structure shown in Figure 4.10. In the analysis of the functioning scenarios, it is of interest to integrate observations or evidence like events in the DBN. If evidence about a component is introduced in the DBN for a state variable $x_i^{(k)}$ or an exogenous variable $u_i^{(k)}$, the inference is correct until the processes are independent. However, if evidence is introduced on a variable, for instance $y_i^{(k+1)}$, and this evidence introduces a dependence between the variables $x_i^{(k+1)}$, then a computing problem appears. This dependence requires the use of specific algorithms. So, it is necessary to be cautious when using DBN and evidence to compute the distributions correctly by considering the right hypothesis.

4.5. Conclusion

In this chapter, DBN are introduced. The chapter shows that DBN can decrease the effects of combinatory explosion when assessing the reliability by the factorization of multi-state systems [WEB 03]. DBN are a representation formalism for MC [WEB 03], HMM and IOHMM. They are well suited for modeling and assessing component reliabilities [WEB 04, BEN 06]. DBN are also well suited to model time-varying non-homogeneous MC [WEB 04]. DBN are also able to model IOHMM by integrating exogenous variables for component degradation modeling [BEN 06].

In addition, DBN are able to model the wearing out or aging of a component. It can also integrate the impact of maintenance actions or the evolution of the component operational conditions. In [WEB 06, MUL 04], these kinds of models were applied to the reliability analysis and prognostic of a system. The DBN model is based on a multi-state probabilistic model of the system and a dynamic reliability model of the components.

DBN are also applied to aiding diagnosis [WEB 08]. In this work, a DBN integrates the reliability models into the probabilistic decision model to diagnose faulty components and to solve possible ambiguity in their isolation. In [VER 08, VER 09], BN or DBN are used as the classifier of faults for the diagnostic.

The works done in [BEN 08] address the modeling of temporally sequenced events. The application concerns the monitoring of the railway line where structural characteristics are taken into account for the diagnosis of degradations (i.e. cracks, fragments, random wears etc.). The problem is solved by an IOHMM that shows the difference between the normal situation and degradations. It has been applied to classify singular points of rail defects [BOU 04, BEN 04, SAM 07, OUK 08].

5

Dynamic Bayesian Networks: Integrating Reliability Computation in the Control System

5.1. Introduction

As discussed in the previous chapters, graphical probabilistic models are applied in risk management, maintenance and diagnosis. Nevertheless, these probabilistic models can also be used in control theory applications. The graphical probabilistic model is implemented online in the closed loop before interruption of system functioning for normal or maintenance actions. The purpose of this chapter is to integrate the probabilistic models into control theory to optimize the control strategy according to failures and their impacts on system reliability.

The control strategy has an impact on the system and its performance during the operation. For instance, modifying a control law according to the faults and failure can warrant the system functioning. Nevertheless, overcharging an actuator to compensate for decreased performance can accelerate the degradation of system components. From a long-term perspective, performance cannot be infinitely compensated for. The greater the compensation, the greater the necessary components' overcharge and the faster the degradation rate.

The models necessary to estimate reliability under operational conditions have to be used online during the operational phase. To do this, it is necessary to use formalism well suited to the estimation and prognosis of the functioning modes induced by events occurring throughout the lifetime of the system. Moreover, control methods are not designed to integrate probabilistic information coming from reliability models. Thus, it is necessary to design methods that are able to integrate the probabilistic knowledge into the control algorithms.

This is an open question and this chapter is dedicated to showing our approach to this paradigm. Obviously, the proposed approach is based on DBN presented previously to estimate the fault and failure impact or context variations on fault-tolerant systems.

5.2. Integrating reliability information into the control

Component reliability and system reliability have been less closely examined in the literature on control theory. [GOK 05] proposed integrating the parameters to increase the life of actuators to reduce the maintenance costs. The method is based on the estimation of time before failure according to the past component use and the modification of the component's functioning state if the estimated remaining lifetime is less than expected. [GOK 06] presented some algorithms for adaptive control. The first algorithm maintains the expected actuator lifetime by adjusting its performance level. The other algorithm offers a compromise between the actuator performance and the expected lifetime according to mission requirements.

[PER 10] proposed a solution based on model predictive control (MPC) strategy that is used to allocate the effort among the redundant actuators by fixing constraints on the actuator degradation. This degradation is computed by cumulating the control inputs. This constraint is integrated in the MPC strategy to protect against the dangerous degradation levels of some critical actuators. This method is not based on reliability computation but integrates the co-variables (control input) that have an impact on the component reliability. The principle is to focus on increasing the component reliability without considering the system reliability.

[GUE 07, GUE 11] focused on defining a structure that combines components to elaborate a system with higher reliability level after a component failure. It is based on a fault-tolerant control whose fundamental principle is to keep the performance levels closer to the performance level defined before the occurrence of failure. Fault tolerance is a control reconfiguration or a restructuring strategy integrating reliability analysis and component costs [GUE 04a, GUE 04b]. From the fault detection and isolation process, the reconfiguration task consists of determining the possible structures that ensure the initial system performance or accepted degraded performances by isolating the faulty components or switching to operating subsystems. For this purpose, an optimal structure is searched for from among all the possible structures [GUE 04a, GUE 05, GUE 06].

[KHE 11] proposed a fault-tolerant control strategy to warrant the system reliability. This new methodology requires adaptation of several reliability models or parameters to integrate them as constraints or conditioning criteria of the control law. The integration of the impact of reliability on the end of mission is a key point of this work.

5.3. Control integrating reliability modeled by DBN

The goal is to define a control strategy for over-actuated systems that allows us to optimally allocate the effort on actuators under the constraint of preserving the system reliability in the normal case or when component failure occurs. To optimize the actuator inputs, it is necessary to have sufficient free degrees in the control law. Clearly, this is the case in over-actuated systems. An over-actuated system is not necessarily a system with redundant components but a system where the control goals can be attained in a different manner.

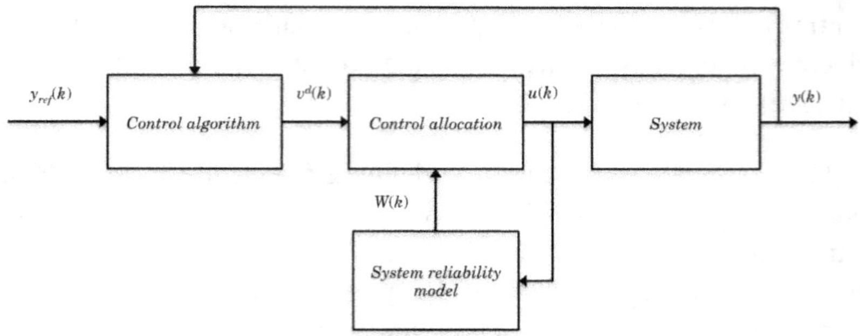

Figure 5.1. *Control structure of an over-actuated system integrating a reliability model*

5.3.1. *Modeling and controlling an over-actuated system*

From a general point of view, an over-actuated system can be considered as a linear system with m actuators and described by the following discrete equation:

$$\begin{cases} \tilde{x}\,(k+1) = A\tilde{x}\,(k) + B_u\tilde{u}\,(k) \\ \tilde{y}\,(k+1) = C\tilde{x}\,(k) \end{cases} \qquad [5.1]$$

with $A \in R^{n\times n}$, $B_u \in R^{n\times m}$ and $C \in R^{p\times n}$ being the state, control and output matrices respectively. $\tilde{x} \in R^n$ is the state vector of the system, $\tilde{u} \in R^m$ is the input control vector and $\tilde{y} \in R^p$ is the system output vector. The condition $rank\,(B_u) = r < m$ characterizes over-actuated systems. Figure 5.1 shows the control principle of an over-actuated system integrating reliability information. The reliability model is used to allocate the control efforts on the actuators.

Matrix B_u can be factorized:

$$B_u = B_v.B \qquad [5.2]$$

with $B_v \in R^{n \times r}$ and $B \in R^{r \times m}$ all of rank r. The system is then modeled by:

$$\begin{cases} \tilde{x}\,(k+1) = A\tilde{x}\,(k) + B_v\tilde{v}\,(k) \\ \tilde{v}\,(k) \quad = \quad B\tilde{u}(k) \\ \tilde{y}\,(k+1) = \quad C\tilde{x}\,(k) \end{cases} \qquad [5.3]$$

with $\tilde{v}\,(k) \in R^r$ representing the whole controlling effort required for the system to function is also called the virtual input vector. Control allocation aims to define the real control inputs of the system $\tilde{u}(k)$ from the expected virtual control input, such as:

$$\begin{array}{l} \tilde{v}^d\,(k) = B\tilde{u}\,(k) \\ \tilde{u}_{min} \leq \tilde{u} \leq \tilde{u}_{max} \end{array} \qquad [5.4]$$

where $\tilde{v}^d\,(k)$ is computed by an algorithm that should satisfy the control and target goals and $\tilde{u}_{min} \leq \tilde{u} \leq \tilde{u}_{max}$ represents the physical limits of the actuators (saturations).

A solution to the allocation problem is given by the resolution of an optimization problem. If no solution exists, an optimal solution is defined with the limits $\tilde{u}(k)$, such that $B\tilde{u}\,(k)$ gives the best approximation of $\tilde{v}^d\,(k)$. The optimal control input can be obtained by minimizing the following criteria:

$$\Psi = \underset{\tilde{u}_{min} \leq \tilde{u} \leq \tilde{u}_{max}}{\arg\min} \; \frac{1}{2} \left\| B\tilde{u}(k) - \tilde{v}^d(k) \right\|_2 \qquad [5.5]$$

$$\tilde{u}\,(k) = \underset{\tilde{u} \in \Psi}{\arg\min} \left\| W(k)(\tilde{u}(k) - \tilde{u}^d(k)) \right\|_2 \qquad [5.6]$$

with Ψ all the possible solutions for the control input $\tilde{u}(k)$, according to the controlling goals and $\tilde{u}^d\,(k)$ being the expected control input. Matrix $W(k) \in R^{m \times m} \succ 0$ allows priority levels to be set for the actuators. $W(k)$ is usually defined as a diagonal matrix:

$$W(k) = diag(w_1(k)w_2(k) \cdots w_i(k) \cdots w_m(k)) \qquad [5.7]$$

5.3.2. *Integrating reliability*

The weighing matrix $W\,(k)$ is considered the key to integrating the actuators' reliability into the control input allocation problem of over-actuated systems. The control problem can be solved in several steps, as shown in Figure 5.2. To maximize system reliability, the weighing matrix $W(k)$ is set from the actuator contributions α_i^k to the system operation:

$$P(\alpha_i^k = 0) \qquad [5.8]$$

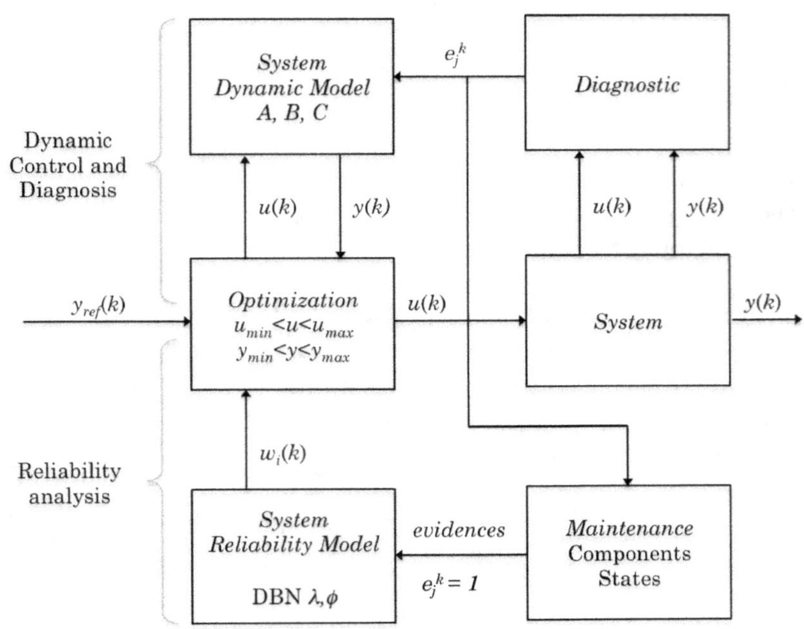

Figure 5.2. *Control framework of an over-actuated system integrating a DBN reliability model*

This contribution depends on the structure function $\varphi(e^k)$, where $e^k = (e_1^k, e_2^k, \ldots, e_m^k)$ allows the calculation of system reliability according to the actuator e_i^k. The actuators are taken into account in the control strategy proportional to their contribution to the system operation. The system state S^k is defined from the structure function $\varphi(e^k)$:

$$P\left(S^k = 0\right) = P(\varphi(e^k) = 0) \qquad [5.9]$$

From the control point of view, the over-actuated system is hypothesised to be in a working state even if some actuators are in a failed state, i.e. $e_j^k = 1$. To satisfy the system goals, the actuators to be used depend on their availability and the structure function φ.

The unavailable actuators $e_j^k = 1$ are isolated by the *Maintenance* function, as shown in Figure 5.2, from the *Diagnostic* function. An available actuator e_i^k is used by the control law if at least one operating scenario exists, i.e. $\varphi(e^k) = 0$, using e_i^k. The probability of using an actuator and satisfying the system objectives is defined by the following conditional probability:

$$P\left(\alpha_i^k = 0 | \varphi(e^k) = 0\right) \qquad [5.10]$$

Variable	Actuator name	Failure rates (10^{-4})
e_1	$iEstrella12$	1.2
e_2	$iEstrella3456$	3.456
e_3	$iSJDSub$	6.3
e_4	$iSJDSpf$	9.5
e_5	$iRelleu$	1
e_6	$iCornella100$	10
e_7	$iCornella50$	5
e_8	$iSJD50$	5
e_9	$iSJD10$	1

Table 5.1. *Failure rates of the actuators*

To integrate the actuators' reliability in the control strategy, the weighing matrix $W(k)$ is estimated online according to the actuators' state given by the *Diagnostic* function. Consequently, if an actuator e_j^k is unavailable, i.e. $e_j^k = 1$, the system can work in the degraded mode because it is over-actuated and the scalar $w_i\,(k)$ of each actuator is defined by the following probability:

$$w_i\,(k) = P\left(\alpha_i^k = 0|\varphi(e^k) = 0, e_j^k = 1\right) \qquad [5.11]$$

The weight $w_i\,(k)$ corresponds to the contribution probability of the actuator e_i^k when the system is working, given the unavailability of some failed actuators. This probability assessment is not only based on the actuator's state of health but also considers the system structure and the availability of other actuators. By using usual reliability modeling, assessing this probability is complex or impossible, given the structure function φ. However, this assessment can easily be realized by the inference mechanism of DBN.

5.4. Application to a drinking water network

The proposed method has been applied to a drinking water network (DWN) in Barcelona, Spain (Figure 5.3). A DWN is typically an over-actuated system because several paths can provide water to customers from various sources. The flow of water on these paths is controlled by actuators such as valves and pumps. The Barcelona DWN is supervised by a two-level monitoring system. At the top level, a control and monitoring system is installed at the control center which is in charge of the whole network, considering operational constraints and customer requirements.

A DWN is a flow network that connects sources of water to the consumers (sink sectors) through pipes. All the DWN parts can fail, i.e. sources, pipes, tanks, connectors, valves and pumps. Thanks to a DBN, all these elements' reliability can be modeled, but the present work focuses on pumps and valves. The other components are considered as perfectly reliable and without any saturation.

The method has been applied in simulation with five sources and one sink, as shown in Figure 5.3. To satisfy the requested demand for each day $d = y_{ref}\,(k)$ ($C100CFE$), at least three sources are required from among the five available sources ($AportLL1$, $AportLL2$, $ApousE12$, $ApousE3456$, and $Aporta$).

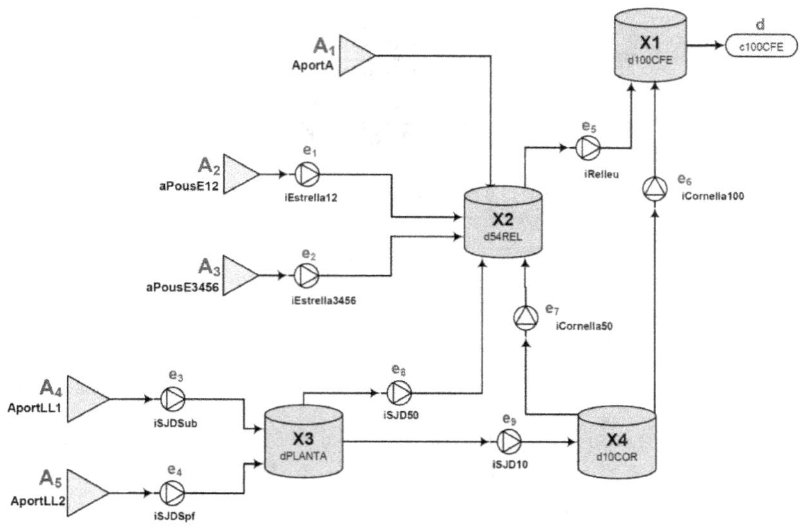

Figure 5.3. *Part of the Barcelona DWN studied*

Variable	Paths
P_1	$\{AportLL2, iSJDSpf, iSJD10, iCornella100\}$
P_2	$\{AportLL1, iSJDSub, iSJD10, iCornella100\}$
P_3	$\{AportLL2, iSJDSpf, iSJD10, iCornella50, iRelleu\}$
P_4	$\{AportLL1, iSJDSub, iSJD10, iCornella50, iRelleu\}$
P_5	$\{AportLL2, iSJDSpf, iSJD50, iRelleu\}$
P_6	$\{AportLL1, iSJDSub, iSJD50, iRelleu\}$
P_7	$\{ApousE12, iEstrella12, iRelleu\}$
P_8	$\{ApousE3456, iEstrella3456, iRelleu\}$
P_9	$\{Aporta, iRelleu\}$

Table 5.2. *Paths linking the sources to the demand point*

5.4.1. *DBN modeling*

All sources are considered to be available. The failure time distributions of all actuators are considered exponential and the failure rates are given in Table 5.1. All paths linking the sources ($AportLL1$, $AportLL2$, $ApousE12$, $ApousE3456$ and $Aporta$) to the demand point ($C100CFE$) are detailed in Table 5.2.

Variable	Actuator weighing
α_{e_1}	WiEstrella12
α_{e_2}	WiEstrella3456
α_{e_3}	WiSJDSub
α_{e_4}	WiSJDSpf
α_{e_5}	WiRelleu
α_{e_6}	WiCornella100
α_{e_7}	WiCornella50
α_{e_8}	WiSJD50
α_{e_9}	WiSJD10

Table 5.3. *Variables Computing the actuator Weighting*

The DBN-based reliability model is shown in Figure 5.4. The P_i variables model the availability of paths linking the sources to the demand. The A_i' variables define the availability of sources through the DWN, according to the availability of the paths. The system availability is defined by the variable $S_{C100CFE}$. The variables computing the weights $w_i(k)$ are defined in Table 5.3.

5.4.2. *Results and discussion*

A simulation is done over 3,600 hours by using Matlab to apply the methodology based on control input optimization and DWN reliability computation. The DBN is based on the Bayes Net Toolbox (BNT), as proposed by Murphy in 2002 [MUR 02].

The results for a particular simulation are illustrated in Figure 5.5. The figure shows the actuator reliability and the DWN reliability. The actuator reliability follows a decreasing exponential curve, whereas the DWN reliability is a bit more complex because it is the combination of the actuator reliability according to the system structure function φ.

Figure 5.6 shows the control inputs applied to the actuators and the weights used for the optimization. The curves $\alpha_i(k)$ show the weight values $w_i(k) = P\left(\alpha_i^k = 0|\varphi(e^k) = 0, e_j^k = 1\right)$ according to the component reliability, the path availability linking sources to demand and the potential component failures or repairs. Each weight $w_i(k)$ is integrated in the trace of the matrix $W(k)$ for the optimization process and the computation of $\tilde{u}(k)$.

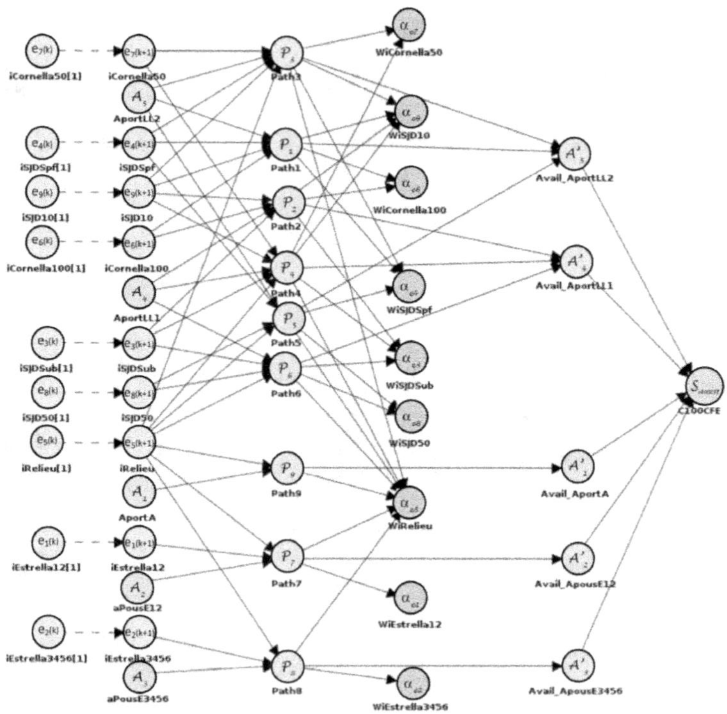

Figure 5.4. *DBN model of the Barcelona DWN*

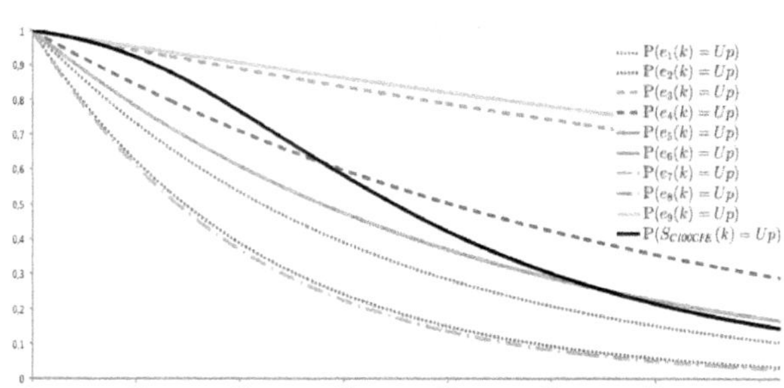

Figure 5.5. *Actuators and DWN reliability*

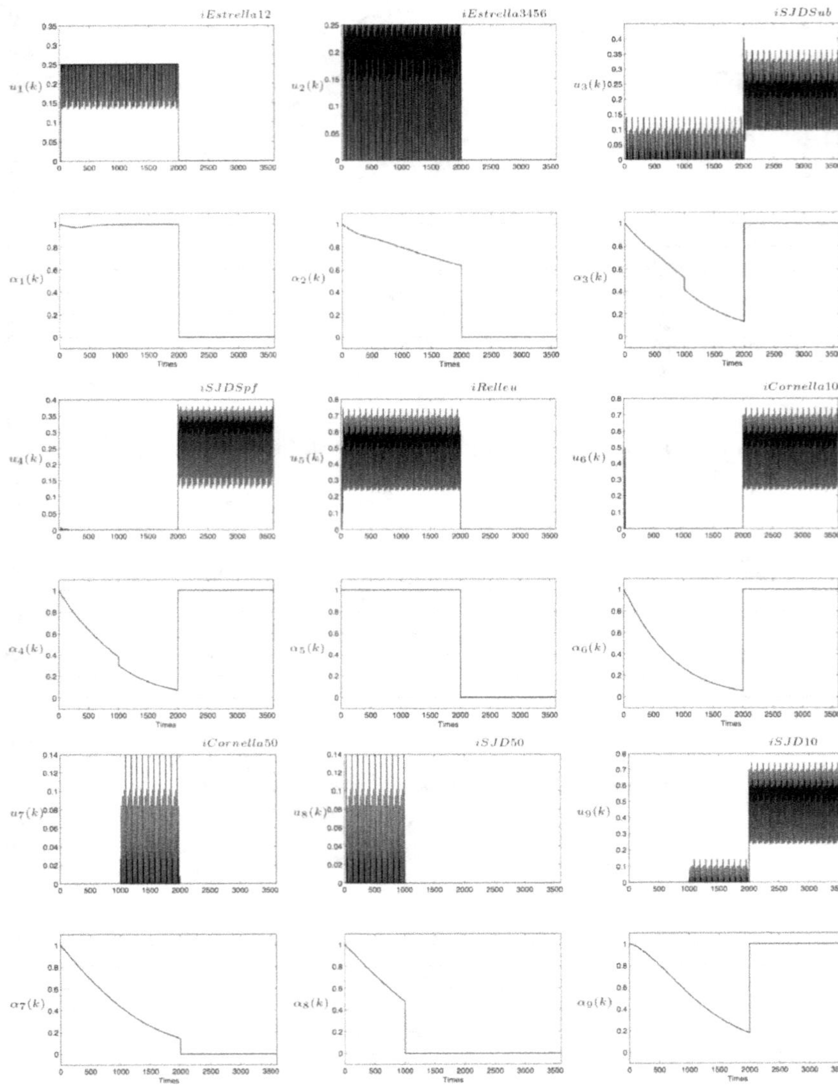

Figure 5.6. *Simulation of control inputs and weights of the Barcelona DWN*

A simulation scenario is divided into three steps:

– Step 1. No failure of the pumping station is considered but source $(Aporta)$ unavailability is simulated by $(k = 0)$ and $(k = 1,000)$. In this case, $ApousE12$ is easily available through path P_7, with components e_5 $(iRelleu)$ and e_1 $(iEstrella12)$ that are the most reliable actuators. Paths P_7 and P_8 are used to satisfy consumer demand. The control inputs $\tilde{u}_1(k)$ and $\tilde{u}_2(k)$ are then used at maximum capacity. However, actuator saturations e_1 and e_2 do not meet the daily peak demand value. It is thus $\tilde{u}_3(k)$ and $\tilde{u}_8(k)$ that are considered to compensate for the saturations of $\tilde{u}_1(k)$ and $\tilde{u}_2(k)$. Component e_5 is mainly considered because this component is more reliable than e_6. The DWN functioning is thus based on links P_6, P_7 and P_8.

– Step 2. A critical failure occurs in e_8 $(iSJD50)$ at $(k = 1,000)$. A *Diagnostic* function is supposed to detect and isolate the failures correctly. The DBN computes the actuator weight $w_i(k)$ by integrating the unavailability of component e_8. In Figure 5.6, the curve of $\alpha_8(k)$ goes to 0. The path P_6 cannot be used to provide water to the consumer. The control strategy switches to path P_4. Actuators e_7 and e_9 that were previously unused are activated. Between $(k = 1000)$ and $(k = 2000)$, the components used are then e_1, e_2, e_3, e_5, e_7 and e_9. The corresponding control inputs are then different from 0.

– Step 3. At $(k = 2000)$, a second critical failure occurs and is simulated on pump e_5 $(iRelleu)$. The control strategy gives priority to e_5 that is not valid and paths P_4, P_7 and P_8 cannot be used. Consequently, the DBN computes the weights $\alpha_1(k)$, $\alpha_2(k)$, $\alpha_5(k)$ and $\alpha_7(k)$ as being equal to 0 and the weight $\alpha_8(k)$. Only the paths containing e_6 $(iCornella100)$, i.e. P_1 and P_2, cannot be used. The system is then reconfigured to work with components e_3, e_4, e_6 and e_9. The DBN computes the weights $\alpha_3(k)$, $\alpha_4(k)$, $\alpha_6(k)$ and $\alpha 9(k)$ as equal to 1 because there is no other valid working scenario.

5.5. Conclusion

A modeling method based on DBN has been proposed in this chapter. It is dedicated to reliability modeling of the system and its actuators to integrate the component states into the control allocation of the system being modeled.

Each step of the system has been formalized. The control process combines a load allocation for all the actuators given the actuators' reliability and their contribution to the system's working state. It is not only the goal to preserve the components' health but also to ensure system availability. This original strategy is applied both in normal conditions and in failure scenarios.

5.6. Acknowledgments

I want to warmly thank my scientific colleagues Prof. Vicenc Puig, Prof. Didier Theilliol, Prof. Fatiha Nejjari and Prof. Ramon Sarrate Estruch, from the Universitat Politécnica de Catalunya (UPC) (Spain) for their contribitions to this chapter.

Conclusion

Despite the maturity of BN's formalism and their application in industrial problems, it is necessary to formalize the procedures or the construction methodologies of the model. If the data are not available for automatic learning of the model, a major effort must be made to focus on the structuring of knowledge and construction of the model because the robustness of the model is based on the quality and rightness of the modeling methodology.

Modeling the functional consequences of failures from structured knowledge

Several modeling methods are proposed in this book to formalize the BN to compute the reliability and impact of degradation of components in the functional architecture of the system. Components and failure modes are then naturally described by multi-state variables, e.g. derived from FMEA or HAZOP analysis, which is difficult to model by other methods.

This book proposes defining the model with different levels of abstraction according to knowledge from functional analysis tools such as structured analysis and design technique, also known as IDEF0. The impact (probabilistically) of failure propagated on the system functions appears in terms of realization of the function or occurrence of failure modes.

The PRM modeling approach is very interesting because it allows the development of models from elementary structures. Thanks to this flexibility, the construction of a model with different levels of abstraction is facilitated. PRM capabilities capitalize knowledge by creating generic classes that need to be instantiated on a particular system.

The inference algorithms in the PRM facilitate the calculation of large models and complex systems. The use of the concept of class allows the definition of a family of objects that share common properties in a probabilistic network: graph, attributes, references and probabilistic relationships. It is a modeling formalism that should become unavoidable in dependability [MED 13, MED 15].

Dynamic modeling system reliability based on the reliability of components from the environment

This book shows that through the factorization of the joint distribution, DBN allows significantly lower complexity in a MC and a setting that is much easier to achieve. DBN extends the use of variations of parameters in time, which broadens the modeling power to the non-homogeneous MC.

Taking into account the dynamic contextual variables, represented by exogenous constraints increases the DBN modeling power structures equivalent to the MSM, HMM and IOHMM which are used in the modeling framework of component reliability. These models correspond naturally to the phenomena encountered in reliability calculation. The states of the components are not observable; only the consequences of the degradation of the component are visible through the failure modes that appear in the functional view of the system.

This type of modeling is applicable to the prognosis of functioning. DBN formalize the relation between the dynamics of component reliability taking into account aging (failure rates varying over time) and the functions of the system. The resulting models are large in size and have been proposed using a BN dynamic object orientation [WEB 06] to facilitate the construction and maintenance of the model.

Synthesis of the control law with the aim of optimizing system reliability based on its sensitivity to actuator failures

This book initially focused on the analyses of reliability of actuators but poses the question in a more general context of control. The book proposes a structured control system incorporating optimization functions and variable probabilistic assessment functions related to the reliability of the components and system. The sensitivity of the system reliability degradation due to actuator loads generated by the control is a key to control allocation. This is the direction that gives the best results. This book is the first to propose this idea, and the increase in interest within the scientific community confirms this direction of research. It is a line of research with high application potential for the coming years [SAL 15].

In general, the contributions of this book tend towards the formalization of probabilistic models with large functional and dysfunctional systems by formalizing views that have an impact on the operating environment and behavior of the system. This addresses the current problems faced by industrialists who must integrate more and more parameters in analyses to meet demands and government safety guidelines.

Bibliography

[ALL 03] ALLEN D., DARWICHE A., "New advances in inference by recursive conditioning", *Proceedings of UAI*, pp. 2–10, 2003.

[ANR 11] Structuring knowledge with object oriented Bayesian nets (SKOOB) project, Report no. ANR Project '07 TLOG 021, ANR, French National Research Agency, 2011.

[ANS 94] ANSELL J., PHILLIPS M., *Practical Methods for Reliability Data Analysis*, Oxford University Press Inc., Oxford, 1994.

[ANT 11] ANTONUCCI A., "The Imprecise Noisy-OR Gate", in *Proceedings of the 14th International Conference on Information Fusion (FUSION 2011)*, Chicago, pp. 709–715, 2011.

[AVE 99] AVEN T., JENSEN U., *Stochastic Models in Reliability, Applications of Mathematics*, Springer-Verlag, New York, 1999.

[BAG 01] BAGDONAVICIUS V., NIKULIN M., "Estimation in degradation models with explanatory variables", *Lifetime Data Analysis*, vol. 7, pp. 85–103, 2001.

[BEN 99] BENGIO Y., "Markovian models for sequential data", *Neural Computing Surveys*, vol. 2, pp. 129–162, 1999.

[BEN 04] BEN SALEM A., BOUILLAUT L., AKNIN P. *et al.*, "Dynamic Bayesian networks for classification of rail defects", in *IEEE 4th International Conference on Intelligent Systems Design and Applications*, Budapest, Hungary, August 26–28 2004.

[BEN 06] BEN SALEM A., MULLER A., WEBER P., "Dynamic Bayesian networks in system reliability analysis", in *6th IFAC Symposium on Fault Detection, Supervision and Safety of Technical Processes*, Beijing, P.R. China, pp. 481–486, August 30 2006.

[BEN 08] BEN SALEM A., Modèles Probabilistes de Séquences Temporelles et Fusion de Décisions. Application à la Classification de Défauts de Rails et à leur Maintenance, Doctoral thesis, Nancy University, University of Lorraine, 2008.

[BEN 13] BENSI M., KIUREGHIAN A.D., STRAUB D., "Efficient Bayesian network modeling of systems", *Reliability Engineering & System Safety*, vol. 112, pp. 200–213, 2013.

[BIN 97] BINDER J., KOLLER D., RUSSELL S. *et al.*, "Adaptive probabilistic networks with hidden variables", *Machine Learning*, vol. 29, no. 2, pp. 213–244, 1997.

[BOB 01] BOBBIO A., PORTINALE L., MINICHINO M. *et al.*, "Improving the analysis of dependable systems by mapping fault trees into Bayesian networks", *Reliability Engineering and System Safety*, vol. 71, no. 3, pp. 249–260, March 2001.

[BOB 03] BOBBIO A., CIANCAMERLA E., FRANCESCHINIS G. *et al.*, "Sequential application of heterogeneous models for the safety analysis of a control system: a case study", *Reliability Engineering and System Safety*, vol. 81, pp. 269–280, 2003.

[BOU 99] BOUTILLIER C., DEAN T., HANKS S., "Decision-theoretic planning: structural assumptions and computational leverage", *Journal of Artificial Intelligence Research*, vol. 11, pp. 1–94, 1999.

[BOU 04] BOUILLAUT L., WEBER P., SALEM A.B. *et al.*, "Use of causal probabilistic networks for the improvement of the maintenance of railway infrastructure", *IEEE International Conference on Systems, Man and Cybernetics*, Hague, Netherlands, October 10–13 2004.

[BOU 05a] BOUDALI H., DUGAN J., "A discrete-time Bayesian network reliability modeling and analysis framework", *Reliability Engineering and System Safety*, vol. 87, pp. 337–349, 2005.

[BOU 05b] BOUDALI H., DUGAN J., "A new Bayesian network approach to solve dynamic fault trees", *IEEE Reliability and Maintainability Symposium.*, pp. 451–456, January 24–27 2005.

[BOU 06] BOUDALI H., DUGAN J., "A continuous-time Bayesian network reliability modeling, and analysis framework", *IEEE Trans on Reliability*, vol. 55, no. 1, pp. 86–97, March 2006.

[BOY 98] BOYEN X., KOLLER D., "Tractable for complex stochastic processes", in *Proceedings of the 14th Annual Conference on Uncertainty in Artificial Intelligence (UAI98)*, Madison, Wisconsin, pp. 33–42, July 1998.

[CEL 06] CELEUX G., CORSET F., LANNOY A. *et al.*, "Designing a Bayesian network for preventive maintenance from expert opinions in a rapid and reliable delay", *Reliability Engineering and System Safety*, vol. 91, no. 7, pp. 849–856, 2006.

[CHE 06] CHEVREAU F., WYBO J., CAUCHOIS D., "Organizing learning processes on risks by using the bowtie representation", *Journal of Hazardous Materials*, vol. 130, no. 3, pp. 276–283, 2006.

[COC 97] COCOZZA-THIVENT C., *Processus stochastiques et fiabilité des systèmes*, Springer, 1997.

[COR 75] CORAZZA M., *Techniques mathématiques de la fiabilité prévisionnelle*, SUP'AERO, Cepadues, 1975.

[COX 55] COX D., "The analysis of non-Markovian stochastic processes by the inclusion of supplementary variables", in *Proceedings of the Cambridge Philosophical Society*, vol. 51, pp. 433–440, 1955.

[DE 92] DE SOUZA E., OCHOA P., "State space exploration in Markov models", *Performance Evaluation Review*, vol. 20, no. 1, pp. 152–166, 1992.

[DE 12] DE ROCQUIGNY E., *Modeling Under Risk and Uncertainty: an Introduction to Statistical, Phenomenological and Computational Methods*, Wiley, 2012.

[DEP 75] DEPARTMENT OF TRANSPORT, Health and safety executives: the Flixborough disaster – The court of enquiry, US Department of Transport, 1975.

[DIE 93] DIEZ F., "Parameter adjustment in Bayes networks, the generalized noisy OR-gate", in *Proceedings of 9th Annual Conference on Uncertainty in Artificial Intelligence (UAI-93)*, Washington, D.C., pp. 99–105, 1993.

[DUG 92] DUGAN J., BAVUSO S., BOYD M., "Dynamic fault-tree for fault-tolerant computer systems", *IEEE Trans Reliability*, vol. 41, no. 3, pp. 363–377, 1992.

[DUV 12] DUVAL C., FALLET-FIDRY G., IUNG B. *et al.*, "A Bayesian network-based integrated risk analysis approach for industrial systems: application to heat sink system and prospects development", in *Proceedings of the Institution of Mechanical Engineers, Part O: Journal of Risk and Reliability*, vol. 226, no. 5, pp. 488–507, 2012.

[FAL 12] FALLET G., AiDR: Eléments pour l'amélioration de la robustesse et la propagation des incertitudes résiduelles, Doctoral thesis, Nancy University, University of Lorraine, 2012.

[FAY 00] FAY A., JARAY J., "A justication of local conditioning in Bayesian networks", *International Journal of Approximate Reasoning*, vol. 24, no. 1, pp. 59–81, 2000.

[GER 00] GERTSBAKH I., *Reliability Theory: with Applications to Preventive Maintenance*, Springer-Verlag, Berlin, 2000.

[GOK 05] GOKDERE L., L.CHIU, KELLER K. *et al.*, "Lifetime control of electromechanical actuators", *Aerospace Conference, 2005 IEEE*, pp. 3523–3531, March 2005.

[GOK 06] GOKDERE L., BOGDANOV A., CHIU S. *et al.*, "Adaptive control of actuator lifetime", *Aerospace Conference, 2006 IEEE*, p. 11, 2006.

[GON 11] GONZALES C., WUILLEMIN P.-H., "PRM inference using Jaffray & Fay's local conditioning", *Theory and Decision*, vol. 71, pp. 33–62, 2011.

[GUE 04a] GUENAB F., JOIN C., PONSART J. *et al.*, "A reliability approach to reconfiguration strategy: application to the IFATIS benchmark problem, session invitee", *2nd IFAC Symposium on System Structure and Control*, Oaxaca, Mexico, December 8–10 2004.

[GUE 04b] GUENAB F., THEILLIOL D., WEBER P. *et al.*, "Fault-tolerant control design based on cost and reliability analysis", *Workshop on Advanced Control and Diagnosis (ACD*, Karlsruhe, Germany, November 17–18 2004.

[GUE 05] GUENAB F., THEILLIOL D., WEBER P. *et al.*, "Fault-tolerant control design based on cost and reliability analysis", *16th IFAC World Congress*, Prague, Czech Republic, July 4–8 2005.

[GUE 06] GUENAB F., THEILLIOL D., WEBER P. *et al.*, "Fault tolerant control system design: a reconfiguration strategy based on reliability analysis under dynamic behavior constraints", in *6th IFAC Symposium on Fault Detection, Supervision and Safety of Technical Processes*, Beijing, China, pp. 1387–1392, 2006.

[GUE 07] GUENAB F., Contribution aux systèmes tolérants aux défauts: synthèse d'une méthode de reconfiguration et/ou de restructuration intégrant la fiabilité des composants., Doctoral Thesis, Nancy University, University of Lorraine, February 20 2007.

[GUE 11] GUENAB F., WEBER P., ZHANG D.T.Y., "Optimal design of fault tolerant control system versus reliability analysis under dynamic behaviour constraints", *International Journal of Systems Science*, vol. 42, no. 1, pp. 219–233, 2011.

[GUY 09] GUYOT B., CONDAMIN L., NAIM P. *et al.*, Validation et représentativité d'un Réseau Bayésien en analyse des risques et sûreté de fonctionnement, Report no. P09-2, Institut de la Maitrise des Risques, 2009.

[HEC 96] HECKERMAN D., "A tutorial on learning with Bayesian networks", Technical Report MSR-TR-95-06, Microsoft Corporation, Washington, USA, 1996.

[HEN 89] HENRION M., "Some practical issues in constructing belief networks", *Uncertainty in Artificial Intelligence*, vol. 3, pp. 213–244,1989.

[HUN 99] HUNG K., VENKATESK S., WEST G., "Layered dynamic probabilistic networks for spatio-temporal modelling", *Intelligent Data Analysis*, vol. 3, pp. 339–361, 1999.

[IEC 06] IEC, NF EN 61 165 – Application des techniques de Markov, Report, IEC, 2006.

[IEC 09] IEC, Risk management – principles and guidelines, Standard, IEC, 2009.

[JAE 02] JAEGER M., "Relational Bayesian networks: a survey", *Electronic Transactions on Artificial Intelligence*, vol. 7, no. 15, 2002.

[JEN 90] JENSEN F., LAURITZEN S., OLESEN K., "Bayesian updating in causal probabilistic networks by local computations", *Computational Statistics Quarterly*, vol. 5, no. 4, pp. 269–282, 1990.

[JEN 96] JENSEN F., *An Introduction to Bayesian Networks*, UCL Press, London, 1996.

[KAN 99] KANG C., GOLAY M., "A Bayesian belief network-based advisory system for operational availability focused diagnosis of complex nuclear power systems", *Expert Systems with Applications*, vol. 17, no. 1, pp. 21–32, 1999.

[KEM 79] KEMENY J., Report of the President's Commission on the Accident at Three Mile Island, Report, Commission on the Accident at TMI, 1979.

[KHA 13] KHAKZADA N., KHANA F., AMYOTTE P., "Dynamic safety analysis of process systems by mapping bowtie into Bayesian network", *Process Safety and Environmental Protection*, vol. 91, pp. 46–53, 2013.

[KHE 11] KHELASSI A., Nouvelle méthodologie de synthèse de lois de commande active tolérante aux fautes garantissant la fiabilité du système., Doctoral thesis, Nancy University, University of Lorraine, 2011.

[KJA 95] KJAERULFF U., "dHugin: a computational system for dynamic time-sliced Bayesian networks", *International Journal of Forecasting*, vol. 11, pp. 89–111, 1995.

[KOL 99] KOLLER D., LERNER U., ANGUELOV D., "A general algorithm for approximate inference and its application to hybrid Bayes nets", *UAI 1999*, pp. 324–333, 1999.

[KOL 00] KOLLER D., LERNER U., *Sampling in Factored Dynamic Systems, Invited Contribution to the Book Sequential Monte Carlo Methods in Practice*, Springer-Verlag, pp. 470–490, 2000.

[KON 80] KONTOLEON J. M., "Reliability determination of a r-Successive-out-of-n:F system", *IEEE Transactions on Reliability*, vol. R-29, no. 5, pp. 437–437, December 1980.

[KÖN 10] KÖNIG J., NORDSTRÖM L., EKSTEDT M., "Probabilistic relational models for assessment of reliability of active distribution management systems", *Probabilistic Methods Applied to Power Systems (PMAPS), 2010 IEEE 11th International Conference*, pp. 454–459, June 2010.

[KRA 98] KRAUSE P., "Learning probabilistic networks", *The Knowledge Engineering Review*, vol. 13, no. 4, pp. 321–351, 1998.

[LAN 07] LANGSETH H., PORTINAL L., "Bayesian networks in reliability", *Reliability Engineering and System Safety*, vol. 92, no. 1, pp. 92–108, January 2007.

[LAN 08] LANGSETH H., "Bayesian Networks in Reliability: the Good, the Bad and the Ugly", in *Advances in Mathematical Modeling for Reliability*, IOS Press, Amsterdam, Netherland, 2008.

[LAU 95] LAURITZEN S., "The EM algorithm for graphical association models with missing data", *Computational Statistics and Data Analysis*, vol. 19, pp. 191–201, 1995.

[LE 14] LE T., CHATELAIN F., BERENGUER C., "Hidden Markov models for diagnostics and prognostics of systems under multiple deterioration modes", *European Safety and Reliability Conference – ESREL 2014*, Wroclaw, Poland. Taylor & Francis – CRC Press/Balkema, pp. 1197–1204, September, 2014.

[LÉG 08a] LÉGER A., DUVAL C., FARRET R. *et al.*, "Modeling of human and organizational impacts for system risk analyses", *9th International Probabilistic Safety Assessment and Management Conference, PSAM 9*, Hong Kong, 2008.

[LÉG 08b] LÉGER A., FARRET R., DUVAL C. *et al.*, "A safety barriers-based approach for the risk analysis of socio-technical systems", in *IFAC World Congress*, Coexo: Korea, South, vol. 17, pp. 6938–6943, 2008.

[LÉG 08c] LÉGER A., LEVRAT E., WEBER P. *et al.*, "Methodology for a probabilistic risk analysis of socio-technical systems", *Insight Journal of INCOSE*, vol. 11, no. 3, pp. 25–26, 2008.

[LÉG 09] LÉGER A., Contribution á la formalisation unifiée des connaissances fonctionnelles et organisationnelles d'un système industriel en vue d'une évaluation quantitative des risques et de l'impact des barrières envisagées, PhD thesis, CRAN, Université Henri Poincaré Nancy, 2009.

[LIS 03] LISNIANSKI A., LEVITIN G., *Multi-State System Reliability: Assessment, Optimization and Application*, World Scientific Pub. Co., 2003.

[MAD 99] MADSEN A., JENSEN F., "LAZY propagation: a junction tree inference algorithm based on lazy inference", *Articial Intelligence*, vol. 113, nos. 1–2, pp. 203–245, 1999.

[MAH 01] MAHADEVAN S., ZHANG R., SMITH N., "Bayesian networks for system reliability reassessment", *Structural Safety*, vol. 23, no. 3, pp. 231–251, 2001.

[MED 11] MEDINA-OLIVA G., Modélisation d'un système de production et de son environnement technique, humain et organisationnel par Réseaux Bayésiens Orientés Objet pour le choix de stratégies de maintenance, PhD thesis, University of Lorraine, Nancy, France, 2011.

[MED 13] MEDINA-OLIVA G., WEBER P., IUNG B., "PRM-based patterns for knowledge formalization of industrial systems to support maintenance strategies assessment", *Reliability Engineering & System Safety*, vol. 116, pp. 38–56, 2013.

[MED 15] MEDINA-OLIVA G., WEBER P., IUNG B., "Industrial system knowledge formalization to aid decision-making in maintenance strategie assessment", *Engineering Applications of Artificial Intelligence*, vol. 37, pp. 343–360, 2015.

[MES 02] MESHKAT L., DUGAN J., ANDREWS J., "Dependability analysis of systems with on-demand and active failure modes, using dynamic fault trees", *IEEE Transactions on Reliability*, vol. 51, no. 2, pp. 240–251, June 2002.

[MIC 09] MICHE E., CHAUMETTE R., Démarche d'évaluation des barrières humaines de sécurité – Omega 20, Report no. DRA-09-103041-06026B, Maîtrise des risques accidentels par les dispositions technologiques et organisationnelles – DRA 77, 2009.

[MOG 12] MOGHADDASS R., ZUO M.J., "A parameter estimation method for a condition-monitored device under multi-state deterioration", *Reliability Engineering & System Safety*, vol. 106, pp. 94–103, 2012.

[MOH 09] MOHAGHEGH Z., KAZEMI R., MOSLEH A., "Incorporating organizational factors into probabilistic risk assessment (PRA) of complex socio-technical systems: a hybrid technique formalization", vol. 94, no. 5, pp. 1000–1018, 2009.

[MUL 04] MULLER A., WEBER P., SALEM A.B., "Process model-based dynamic Bayesian networks for prognostic", *IEEE 4th International Conference on Intelligent Systems Design and Applications*, Budapest, Hungary, August 26–28 2004.

[MUN 07] MUNTEANU P., SUHNER C.C.M., WEBER P., Réseaux Bayésiens et retour d'expérience en sûreté de fonctionnement (Etat de l'amélioration – Adaptation spécifique aux EPSdF – Elaboration de recommandations méthodologiques, Report no. IMdR P04-7, Institut pour la Maitrise des Risques, 2007.

[MUR 02] MURPHY K.P., Dynamic Bayesian networks: representation, inference and learning, PhD thesis, 2002.

[ONI 01] ONIKO A., DRUZDZEL M.J., WASYLUK H., "Learning Bayesian network parameters from small data sets: application of noisy-OR gates", *International Journal of Approximate Reasoning*, vol. 27, no. 2, pp. 165–182, 2001.

[OUK 08] OUKHELLOU L., BOUILLAUT L., COME E. *et al.*, "Combined used of sensor data and strucural information resumed by Bayesian network, application to a railway diagnosis-aid scheme", *Transportation Research Part C*, vol. 16, pp. 755–767, 2008.

[PAP 03] PAPAZOGLOU I., BELLAMY L., HALE A. *et al.*, "I-Risk: development of an integrated technical and management risk methodology for chemical installations", *Journal of Loss Prevention in the Process Industries*, vol. 16, no. 6, pp. 575–591, 2003.

[PAT 96] PATÉ-CORNELL M., MURPHY D., "Human and management factors in probabilistic risk analysis: the SAM approach and observations from recent applications", *Reliability Engineering & System Safety*, vol. 53, no. 2, pp. 115–126, 1996.

[PEA 88] PEARL J., *Probabilistic Reasoning in Intelligent Systems: Networks of Plausible Inference*, Morgan Kaufmann Publishers Inc., San Francisco, CA, USA, 1988.

[PEO 91] PEOT M., SHACHTER R., "Fusion and propagation with multiple observations in belief networks", *Articial Intelligence*, vol. 48, pp. 299–318, 1991.

[PER 10] PEREIRA E.B., GALVÃO R.K.H., YONEYAMA T., "Model predictive control using prognosis and health monitoring of actuators", *Industrial Electronics (ISIE), 2010 IEEE International Symposium on*, pp. 237–243, July 2010.

[PLO 04] PLOT E., Guide de la prise en compte du facteur humain dans l'évaluation des systèmes de gestion des risques majeurs, Présentation du référentiel MIRIAM et des protocoles ATOS, Report, 2004.

[POR 10] PORTINALE L., RAITERI D., MONTANI S., "Supporting reliability engineers in exploiting the power of dynamic Bayesian networks", *International Journal of Approximate Reasoning*, vol. 51, no. 2, pp. 179–195, 2010.

[POU 09] POURE P., WEBER P., THEILLIOL D. *et al.*, "Fault tolerant control of a three-phase three-wire shunt active filter system based on reliability analysis", *Electric Power Systems Research*, vol. 2, no. 79, pp. 325–334, February 2009.

[RAB 89] RABINET L., "A tutorial on hidden Markov models and selected applications in speech recognition", *Proceedings of the IEEE*, vol. 77, no. 2, February 1989.

[ROB 13] ROBLÈS B., AVILA M., DUCULTY F. *et al.*, "Evaluation of minimal data size by using entropy in a HMM maintenance manufacturing use", *MIM'2013 Manufacturing Modelling, Management and Control*, St. Petersburg, Russia, 2013.

[SAL 15] SALAZAR J., WEBER P., SARRATE R. *et al.*, "MPC design based on a DBN reliability model: application to drinking water networks", *9th IFAC Symposium on Fault Detection, Supervision and Safety for Technical Processes, SAFEPROCESS'2015, Special Session on: Applications of Probabilistic Graphical Models in Dependability, Diagnosis and Prognosis*, Paris, France, 2015.

[SAM 07] SAMÉ A., BOUILLAUT L., AKNIN P. *et al.*, "Réseaux bayésiens dynamiques à variable exogène continue pour la classification des points singuliers d'une voie ferrée", *Revue des Sciences et Technologies de l'Information – Série RIA: Revue d'Intelligence Artificielle*, vol. 21, no. 3, pp. 353–370, June 2007.

[SEV 82] SEVESO, Original Seveso directive 82/501/EEC (SEVESO I), European Commission, 1982.

[SHA 96] SHAFER G., *Probabilistic Expert Systems*, SIAM, 1996.

[SHU 10] SHUBIN S., ZHIQIANG C., SHUDONG S. *et al.*, "Integrated importance measures of multi-state systems under uncertainty", *Computers & Industrial Engineering*, vol. 59, no. 4, pp. 921–928, 2010.

[SIM 07] SIMON C., WEBER P., LEVRAT E., "Bayesian networks and evidence theory to model complex systems reliability", *Journal of Computers*, vol. 2, no. 1, pp. 33–43, 2007.

[SIM 08] SIMON C., WEBER P., EVSUKOFF A., "Bayesian networks inference algorithm to implement Dempster Shafer theory in reliability analysis", *Reliability Engineering and System Safety*, vol. 93, no. 7, pp. 950–963, 2008.

[SIN 95] SINGPURWALLA N.D., "Survival in dynamic environments", *Statistical Science*, vol. 10, no. 1, pp. 86–103, 1995.

[SOM 10] SOMMESTAD T., EKSTEDT M., JOHNSON P., "A probabilistic relational model for security risk analysis", *Computers & Security*, vol. 29, no. 6, pp. 659–679, 2010.

[SRI 93] SRINIVAS S., "A generalization of the noisy-OR model", in *Proceedings of 9th Annual Conference on Uncertainty in Artificial Intelligence (UAI-93)*, Washington, D.C., pp. 208–215, 1993.

[SVE 02] SVEDUNG I., RASMUSSEN J., "Graphic representation of accident scenarios: mapping system structure and the causation of accidents", *Safety Science*, vol. 40, no. 5, pp. 397–417, 2002.

[TOR 98] TORRES-TOLEDANO J., SUCAR L., "Bayesian networks for reliability of complex systems", in COELHO H. (ed.), *Progress in Artificial Intelligence, IBERAMIA'98*, Springer-Verlag, pp. 195–206, 1998.

[VER 08] VERRON S., WEBER P., THEILLIOL D. *et al.*, "Using Bayesian networks for decision in the simultaneous faults case", *6th Workshop on Advanced Control and Diagnosis*, Coventry: UK, November 27–28 2008.

[VER 09] VERRON S., WEBER P., THEILLIOL D. *et al.*, "Decision with Bayesian network in the concurrent faults event", *7th IFAC Symposium on Fault Detection, Supervision and Safety of Technical Processes*, Barcelona, Spain, June 30–July 3 2009.

[VIL 88] VILLEMEUR A., *Sûreté de Fonctionnement des Systèmes Industriels Fiabilité – Facteurs Humains Informatisation*, EOF, 1988.

[WEB 01] WEBER P., SUHNER M., IUNG B., "System approach-based Bayesian network to aid maintenance of manufacturing process", in *6th IFAC Symposium on Cost Oriented Automation, Low Cost Automation*, Berlin, Germany, pp. 33–39, October 8–9 2001.

[WEB 02] WEBER P., "Dynamic Bayesian networks model to estimate process availability", *8th International Conference Quality, Reliability, Maintenance, CCF'02*, Sinaia, Romania, 2002.

[WEB 03] WEBER P., JOUFFE L., "Reliability modeling with dynamic Bayesian networks", *SAFEPROCESS'03*, June 9–11 2003.

[WEB 04] WEBER P., MUNTEANU P., JOUFFE L., "Dynamic Bayesian networks modeling the dependability of systems with degradations and exogenous constraints", *11th IFAC Symposium on Information Control Problems in Manufacturing (INCOM'04)*, Salvador-Bahia, Brazil, April 5–7 2004.

[WEB 06] WEBER P., JOUFFE L., "Complex system reliability modeling with dynamic object oriented Bayesian networks (DOOBN)", *Reliability Engineering and System Safety*, vol. 91, no. 2, pp. 149–162, February 2006.

[WEB 08] WEBER P., POURE P., THEILLIOL D. *et al.*, "Design of hardware fault tolerant control architecture for Wind Energy Conversion System with DFIG based on reliability analysis", *2008 ISIE Symposium on Industrial Electronics*, pp. 2323–2328, June 2008.

[WEB 10] WEBER P., SIMON C., THEILLIOL D., "Reconfiguration of over-actuated consecutive-k-out-of-n: F systems based on Bayesian network reliability model", *8th Workshop on Advanced Control and Diagnosis*, Ferrara, Italy, November 2010.

[WEB 11] WEBER P., SIMON C., THEILLIOL D. *et al.*, "Control allocation of k-out-of-n systems based on Bayesian network reliability model: application to a drinking water network", *ESREL 2011-Annual Conference*, Troyes, France, September 2011.

[WEB 12a] WEBER P., BECKER F., MATHIAS A. *et al.*, "Reliability analysis of fault tolerant wind energy conversion system with doubly fed induction generator", *Intelligent Robotics and Applications: 5th International Conference, Montreal, 3–5 October, 2012, Proceedings, Part I*, Springer Berlin Heidelberg, pp. 483–492, 2012.

[WEB 12b] WEBER P., MEDINA-OLIVA G., SIMON C. *et al.*, "Overview on Bayesian networks applications for dependability, risk analysis and maintenance areas", *Engineering Applications of Artificial Intelligence*, vol. 25, no. 4, pp. 671–682, June 2012.

[WEL 00] WELCH R., THELEN T., "Dynamic reliability analysis in an operational context: the Bayesian network perspective", in SMIDTS C., DEVOOGHT J., LABEAU P. (eds.), *Dynamic Reliability: Future Directions*, Elsevier, Maryland, USA, 2000.

[WUI 12] WUILLEMIN P.-H., TORTI L., "Structured probabilistic inference", *International Journal of Approximate Reasoning*, vol. 53, no. 7, pp. 946–968, 2012.

[YAM 03] YAM R.C., ZUO M.J., ZHANG Y.L., "A method for evaluation of reliability indices for repairable circular consecutive-k-out-of-n:F systems", *Reliability Engineering & System Safety*, vol. 79, no. 1, pp. 1–9, 2003.

Index

N, O, P, R

S, T, W

Other titles from ISTE in Systems and Industrial Engineering – Robotics

2016

ANDRÉ Michel, SAMARAS Zissis
Energy and Environment
(Research for Innovative Transports Set - Volume 1)

AUBRY Jean-François, BRINZEI Nicolae, MAZOUNI Mohammed-Habib
Systems Dependability Assessment: Benefits of Petri Net Models (Systems Dependability Assessment Set - Volume 1)

BLANQUART Corinne, CLAUSEN Uwe, JACOB Bernard
Towards Innovative Freight and Logistics (Research for Innovative Transports Set - Volume 2)

COHEN Simon, YANNIS George
Traffic Management (Research for Innovative Transports Set - Volume 3)

MARÉ Jean-Charles
Aerospace Actuators 1: Needs, Reliability and Hydraulic Power Solutions

TORRENTI Jean-Michel, LA TORRE Francesca
Materials and Infrastructures 1 (Research for Innovative Transports Set - Volume 5A)

Materials and Infrastructures 2 (Research for Innovative Transports Set - Volume 5B)

YANNIS George, COHEN Simon
Traffic Safety (Research for Innovative Transports Set - Volume 4)

2015

AUBRY Jean-François, BRINZEI Nicolae
Systems Dependability Assessment: Modeling with Graphs and Finite State Automata

BOULANGER Jean-Louis
CENELEC 50128 and IEC 62279 Standards

BRIFFAUT Jean-Pierre
E-Enabled Operations Management

MISSIKOFF Michele, CANDUCCI Massimo, MAIDEN Neil
Enterprise Innovation

2014

CHETTO Maryline
Real-time Systems Scheduling
Volume 1 – Fundamentals
Volume 2 – Focuses

DAVIM J. Paulo
Machinability of Advanced Materials

ESTAMPE Dominique
Supply Chain Performance and Evaluation Models

FAVRE Bernard
Introduction to Sustainable Transports

GAUTHIER Michaël, ANDREFF Nicolas, DOMBRE Etienne
Intracorporeal Robotics: From Milliscale to Nanoscale

MICOUIN Patrice
Model Based Systems Engineering: Fundamentals and Methods

MILLOT Patrick
Designing Human–Machine Cooperation Systems

NI Zhenjiang, PACORET Céline, BENOSMAN Ryad, REGNIER Stéphane
Haptic Feedback Teleoperation of Optical Tweezers

OUSTALOUP Alain
Diversity and Non-integer Differentiation for System Dynamics

REZG Nidhal, DELLAGI Sofien, KHATAD Abdelhakim
Joint Optimization of Maintenance and Production Policies

STEFANOIU Dan, BORNE Pierre, POPESCU Dumitru, FILIP Florin Gh., EL KAMEL Abdelkader
Optimization in Engineering Sciences: Metaheuristics, Stochastic Methods andDecision Support

2013

ALAZARD Daniel
Reverse Engineering in Control Design

ARIOUI Hichem, NEHAOUA Lamri
Driving Simulation

CHADLI Mohammed, COPPIER Hervé
Command-control for Real-time Systems

DAAFOUZ Jamal, TARBOURIECH Sophie, SIGALOTTI Mario
Hybrid Systems with Constraints

FEYEL Philippe
Loop-shaping Robust Control

FLAUS Jean-Marie
Risk Analysis: Socio-technical and Industrial Systems

FRIBOURG Laurent, SOULAT Romain
Control of Switching Systems by Invariance Analysis: Application to Power Electronics

GROSSARD Mathieu, REGNIER Stéphane, CHAILLET Nicolas
Flexible Robotics: Applications to Multiscale Manipulations

GRUNN Emmanuel, PHAM Anh Tuan
Modeling of Complex Systems: Application to Aeronautical Dynamics

HABIB Maki K., DAVIM J. Paulo
Interdisciplinary Mechatronics: Engineering Science and Research Development

HAMMADI Slim, KSOURI Mekki
Multimodal Transport Systems

JARBOUI Bassem, SIARRY Patrick, TEGHEM Jacques
Metaheuristics for Production Scheduling

KIRILLOV Oleg N., PELINOVSKY Dmitry E.
Nonlinear Physical Systems

LE Vu Tuan Hieu, STOICA Cristina, ALAMO Teodoro,
CAMACHO Eduardo F., DUMUR Didier
Zonotopes: From Guaranteed State-estimation to Control

MACHADO Carolina, DAVIM J. Paulo
Management and Engineering Innovation

MORANA Joëlle
Sustainable Supply Chain Management

SANDOU Guillaume
Metaheuristic Optimization for the Design of Automatic Control Laws

STOICAN Florin, OLARU Sorin
Set-theoretic Fault Detection in Multisensor Systems

2012

AÏT-KADI Daoud, CHOUINARD Marc, MARCOTTE Suzanne, RIOPEL Diane
Sustainable Reverse Logistics Network: Engineering and Management

BORNE Pierre, POPESCU Dumitru, FILIP Florin G., STEFANOIU Dan
Optimization in Engineering Sciences: Exact Methods

CHADLI Mohammed, BORNE Pierre
Multiple Models Approach in Automation: Takagi-Sugeno Fuzzy Systems

DAVIM J.Paulo
Lasers in Manufacturing

DECLERCK Philippe
Discrete Event Systems in Dioid Algebra and Conventional Algebra

DOUMIATI Moustapha, CHARARA Ali, VICTORINO Alessandro, LECHNER Daniel
Vehicle Dynamics Estimation using Kalman Filtering: Experimental Validation

GUERRERO José A, LOZANO Rogelio
Flight Formation Control

HAMMADI Slim, KSOURI Mekki
Advanced Mobility and Transport Engineering

MAILLARD Pierre
Competitive Quality Strategies

MATTA Nada, VANDENBOOMGAERDE Yves, ARLAT Jean
Supervision and Safety of Complex Systems

POLER Raul *et al.*
Intelligent Non-hierarchical Manufacturing Networks

TROCCAZ Jocelyne
Medical Robotics

YALAOUI Alice, CHEHADE Hicham, YALAOUI Farouk, AMODEO Lionel
Optimization of Logistics

ZELM Martin *et al.*
Enterprise Interoperability –I-EASA12 Proceedings

2011

CANTOT Pascal, LUZEAUX Dominique
Simulation and Modeling of Systems of Systems

DAVIM J. Paulo
Mechatronics

DAVIM J. Paulo
Wood Machining

GROUS Ammar
Applied Metrology for Manufacturing Engineering

KOLSKI Christophe
Human–Computer Interactions in Transport

LUZEAUX Dominique, RUAULT Jean-René, WIPPLER Jean-Luc
Complex Systems and Systems of Systems Engineering

ZELM Martin, *et al.*
Enterprise Interoperability: IWEI2011 Proceedings

2010

BOTTA-GENOULAZ Valérie, CAMPAGNE Jean-Pierre, LLERENA Daniel, PELLEGRIN Claude
Supply Chain Performance / Collaboration, Alignement and Coordination

BOURLÈS Henri, GODFREY K.C. Kwan
Linear Systems

BOURRIERES Jean-Paul
Proceedings of CEISIE'09

CHAILLET Nicolas, REGNIER Stéphane
Microrobotics for Micromanipulation

DAVIM J. Paulo
Sustainable Manufacturing

GIORDANO Max, MATHIEU Luc, VILLENEUVE François
Product Life-Cycle Management / Geometric Variations

LOZANO Rogelio
Unmanned Aerial Vehicles / Embedded Control

LUZEAUX Dominique, RUAULT Jean-René
Systems of Systems

VILLENEUVE François, MATHIEU Luc
Geometric Tolerancing of Products

2009

DIAZ Michel
Petri Nets / Fundamental Models, Verification and Applications

OZEL Tugrul, DAVIM J. Paulo
Intelligent Machining

PITRAT Jacques
Artificial Beings

2008

ARTIGUES Christian, DEMASSEY Sophie, NERON Emmanuel
Resources–Constrained Project Scheduling

BILLAUT Jean-Charles, MOUKRIM Aziz, SANLAVILLE Eric
Flexibility and Robustness in Scheduling

DOCHAIN Denis
Bioprocess Control

LOPEZ Pierre, ROUBELLAT François
Production Scheduling

THIERRY Caroline, THOMAS André, BEL Gérard
Supply Chain Simulation and Management

2007

DE LARMINAT Philippe
Analysis and Control of Linear Systems

DOMBRE Etienne, KHALIL Wisama
Robot Manipulators

LAMNABHI Françoise *et al.*
Taming Heterogeneity and Complexity of Embedded Control

LIMNIOS Nikolaos
Fault Trees

2006

FRENCH COLLEGE OF METROLOGY
Metrology in Industry

NAJIM Kaddour
Control of Continuous Linear Systems

www.ingramcontent.com/pod-product-compliance
Lightning Source LLC
LaVergne TN
LVHW022056200926
848481LV00003B/932
9781848219922